8th Offshore Energy and Storage Symposium (OSES 2024)

Journal of Physics: Conference Series Volume 2929

New Bedford, Massachusetts, USA
10-12 July 2024

ISBN: 979-8-3313-1193-3
ISSN: 1742-6588

Printed from e-media with permission by:

Curran Associates, Inc.
57 Morehouse Lane
Red Hook, NY 12571

Institute of Physics
Dirac House, Temple Back
Bristol BS1 6BE UK

Phone: 44 1 17 929 7481
Fax: 44 1 17 920 0979

techtracking@iop.org

Additional copies of this publication are available from:

Curran Associates, Inc.
57 Morehouse Lane
Red Hook, NY 12571 USA
Phone: 845-758-0400
Fax: 845-758-2633
Email: curran@proceedings.com
Web: www.proceedings.com

TABLE OF CONTENTS

Preface

The connections between people, energy, and our environment has never been more clear. Global conflicts and socioeconomic divides have further compounded the strain on our critical environmental priorities. Clean energy systems may currently be more expensive than conventional technologies using fossil fuels, but they remain essential to address climate change and energy security. For these systems to become solutions they must be financially viable.

The 8th Offshore Energy & Storage Symposium was held between 10 – 12 July, 2024 in New Bedford, Massachusetts on America's East Coast. The event brought together researchers, industry players and policy makers dedicated to driving the development and growth of offshore renewables and energy storage. This event followed a number of successful OSES conferences held in Windsor, Canada; Edinburgh, Scotland; Valletta, Malta; Cape Cod, USA; Ningbo, China; Brest, France, and St. Julian's, Malta.

The OSES2024 proceedings include 13 papers covering a range of topics, including

- adiabatic compressed air energy storage
- subsea hydro-pneumatic energy storage
- data-driven contract negotiation for the sale of electricity from wind farms integrating storage
- load factor analysis of wind turbines
- electrification of maritime transport using batteries
- renewable energy backup alternatives for data centres
- modelling an island grid with a high penetration of renewables
- finite element analysis of dynamic subsea cables

All papers followed a double anonymous review by experts in the field. We would like to express our sincere gratitude to all reviewers for the throughout reviewer and to all authors for completing the paper revisions in a timely manner. Finally we would also like to thank School for Marine Science & Technology [SMAST] of the University of Massachusetts Dartmouth for hosting the OSES2024 event in New Bedford, MA.

Rupp Carriveau, University of Windsor
Seamus Garvey, University of Nottingham
Tonio Sant, University of Malta

Journal of Physics: Conference Series **2929** (2024) 011001 doi:10.1088/1742-6596/2929/1/011001

Conference Committees

Organising Committee

Prof. Rupp Carriveau, University of Windsor, Canada
Prof. Seamus Garvey, University of Nottingham, UK
Prof. Tonio Sant, University of Malta, Malta
Dipl.-Phys. Jochen Bard, Fraunhofer IEE, Germany
Prof. Kevin Stokesbury, University of Massachusetts Dartmouth, USA
Prof. Christopher Niezrecki, University of Massachusetts Lowell, USA
Bill Hurley, Hurley Maritime Consulting, USA
Sydney Carreiro, University of Massachusetts Dartmouth, USA
Naomi Pelkey, University of Windsor, Canada

Technical Committee

Prof. Rupp Carriveau, University of Windsor, Canada
Prof. Seamus Garvey, University of Nottingham, UK
Prof. Tonio Sant, University of Malta, Malta
Dipl.-Phys. Jochen Bard, Fraunhofer IEE, Germany
Dr. Marianne Rodgers, Wind Energy Institute of Canada, Canada
Prof. Zhiwen Wang, Dalian Maritime University, China
Prof. Sukanta Basu, University at Albany, USA
Prof. Jonathan Miles, James Madison University, USA
Dr. Shawn Sheng, National Renewable Energy Laboratory, USA
Prof. Alex Slocum, Massachusetts Institute of Technology, USA
Prof. Yi Guo, Technical University of Denmark, Denmark
Prof. Christopher Niezrecki, University of Massachusetts Lowell, USA
Dr. Jacqueline Stagner, University of Windsor, Canada
Prof. Matt Davison, Western University Canada, Canada
Bill Hurley, Hurley Maritime Consulting, USA
Prof. Curran Crawford, University of Victoria, Canada
Prof. David Ting, University of Windsor, Canada
Prof. Eric Loth, University of Virginia, USA

Peer Review Statement

All papers published in this volume have been reviewed through processes administered by the Editors. Reviews were conducted by expert referees to the professional and scientific standards expected of a proceedings journal published by IOP Publishing.

- **Type of peer review:** Double Anonymous
- **Conference submission management system:** None
- **Number of submissions received:** 13
- **Number of submissions sent for review:** 13
- **Number of submissions accepted:** 13
- **Acceptance Rate (Submissions Accepted / Submissions Received × 100):** 100
- **Average number of reviews per paper:** 2
- **Total number of reviewers involved:** 15
- **Contact person for queries:**
  Name: Tonio Sant
  Email: tonio.sant@um.edu.mt
  Affiliation: University of Malta

Investigating the thermal performance of a subsea liquid piston air compressor pipeline operating at very high pressure ratios

Charise Cutajar[1*], Tonio Sant[1] and Luke Aquilina[1]

[1] Department of Mechanical Engineering, University of Malta, Msida, Malta

*E-mail: charise.cutajar@um.edu.mt

Abstract. Liquid piston air compression is presently gaining significant interest as a means for long duration energy storage at large scale. However, various aspects attributed to the thermal performance of liquid piston air compressors at high operating pressure ratios remain yet unanswered. This paper presents a transient thermal analysis of a novel, subsea liquid piston air compressor pipeline operating at pressure ratios up to 200. A thermal model implementing different empirical correlations for the convective heat transfer phenomena is presented. The study reveals a major knowledge gap in terms of Nusselt number correlations available to define the internal heat transfer coefficients explicit to liquid piston pipelines. Results from a parametric study suggest that the internal heat transfer inside a liquid piston having a very large length-to-diameter ratio was characterised by natural convection effects. Furthermore, it was found that to attain a thermal efficiency higher than 0.90, the power density had to be limited to not more that 0.22 kW/m^3, with a compression time required per stroke being at least 25minutes.

1. Introduction

Air compression has been of critical relevance in various utilities ranging from refrigeration and air conditioning, to pneumatic control systems and energy storage applications. The latter purpose has seen an upsurge in recent years due to an unprecedent growing share of renewables in the electricity generation, which in turn, present unique challenges to the security of power supply. Indeed, compressed air energy storage (CAES) is presently the second most mature mechanical energy storage system (ESS) following pumped hydro storage (PHS) [1].

CAES systems are increasingly being favoured due to the ability to meet long duration energy storage requirements on a multi-Megawatt-Hour scale. The most recent variation of CAES, which is also of most relevance to the present research, is the isothermal (I-CAES) configuration which adopts a liquid piston (LP) approach. In other words, the excess renewable energy (RE) is utilised to drive a pump that supplies a flow of liquid (such as oil [2], water [3] or seawater [4]) into an accumulator to eventually compress air. When energy recovery is required, the air inside the accumulator is allowed to expand, expelling the liquid to drive a turbine and generate electricity, when and as required. In the conventional and simplest form of I-CAES, the air used to store the energy does not flow out of the accumulator, thus the system operates in a closed-cycle (CI-CAES) configuration [2-4]. Under conditions enabling high heat exchange between the air undergoing compression/expansion and the surrounding, I-CAES systems can approach high thermal efficiencies, closely approaching 100%. This is especially the case for offshore I-CAES where the accumulator is situated underwater [4]. Consequently, the power consumption per unit of stored energy is significantly lower as compared to alternative CAES systems following a polytropic compression/expansion [5]. One major disadvantage of CI-CAES is, however, the limited energy storage densities (ESD). The latter arises due to limited pressure ratios and limited volumetric capacity, whereby the liquid occupies a substantial capacity of the accumulator. In recent years, research on improving the performance of CI-CAES systems has shifted towards revolutionising the CI-CAES to operate in open-cycle isothermal (OI-CAES) configuration. The idea behind the OI-CAES system is to improve the ESD of the system without compromising the high thermal efficiencies. A higher ESD is achieved by operating at high pressure ratios and utilising a series of LPs to compress the air, which is then stored in a dedicated compressed air chamber (CAS). Doing so, however, has proven to be a major challenge since the high operating pressure ratios boost the ESD at the expense of lower thermal efficiencies. Consequently, there has been an augmented urge to understand better the behaviour of liquid piston air compressors (LPACs) and enhance the near-isothermal conditions.

A number of research works [6-9] have indeed attempted to provide better insight on the thermal performance of LPACs in OI-CAES units by adopting different thermal models. However, a number of uncertainties attributable to the thermal performance of LPACs remain yet unanswered. Indeed, one common remark made by all authors in [6-9] was that of the complexity and lack of standardised approach to adequately define the heat transfer coefficient (HTC) values in LPACs. Apart from the fact that the amount of published literature is also somewhat limited, the studies reviewed [6-9] have only considered small-scale LPAC cylinders, operating in the open atmosphere and with pressure ratios limited to a maximum of 100 [7]. Detailed transient thermal analyses of large-scale LPACs operating at higher pressure ratios and in a subsea environment are lacking. Situating the LPACs and ESS subsea minimises space utilisation restrictions. Consequently, long subsea pipelines (rather than cylinders) can be utilised for LPAC applications in OI-CAES systems to increase the length-to-diameter (L/D_o) ratio and augment the surface area for better heat exchange. More importantly, the subsea environment can be exploited as an excellent heat sink/source, thus enabling the thermodynamic compression/expansion process to attain higher efficiencies as compared to land-based OI-CAES systems.

The present study attempts to provide further insight on the expected thermal performance of a novel, subsea liquid piston air compressor pipeline (LPACP). The LPACP is to be operated as an air compressor to advance an offshore OI-CAES system. The paper presents a parametric study based on a transient thermal analysis of a LPACP designed to operate at a high pressure ratio per stroke, up to 200, equating to double the pressure ratios currently reported in literature [7]. The mathematical model and theoretical assumptions on which the investigation is based are outlined in Section 2. The key outcomes obtained are set forth in Section 3, followed by the conclusions drawn thereof, in Section 4.

2. Mathematical model

2.1 Mass and energy conservation

A transient thermal model was developed in Python© [10] to simulate the thermal behaviour of a subsea LPACP across a compression stroke with high pressure ratios. The steel LPACP geometry and corresponding dimensions are presented in Figure 1. The dimensions, including the steel wall thickness, were derived from an in-house developed program [11] based on pressure vessel guidelines and pipeline codes [12-14], which had been developed to drive material reductions in offshore pipeline applications. For the purpose of the study, no thermal insulations were considered. Yet, a 3 mm internal corrosion allowance was considered in accordance with [13]. The mathematical model presented assumes:

i. air as an ideal gas;
ii. air in the LPACP as a lumped mass with no temperature variations along the length of the vessels;
iii. a perfectly vertical air-seawater interface in the horizontal LPACP, as shown in Figure 1;
iv. a single and homogenous HTC at the air-seawater interface. Therefore, variations due to thermal gradients across the interface are not accounted for;
v. negligble frictional heating;
vi. that the LPACP is in the new, deployment (hence smooth surface) condition. Thus, the sensitivity of the thermal performance to a change in surface roughness is not incorporated in the present work.

Furthermore, the model was specifically developed to investigate the transient thermal effects during the compression process only. The air expansion procedure was beyond the scope of the present work. The compression stroke was always assumed to start from atmospheric pressure, with the mass of air already entrapped in the LPACP and occupying the entire volumetric capacity of the pipeline.

Calculations were based on the fundamental principles of mass conservation and energy conservation. Considering the LPACP as an isolated and closed system during compression, the air mass flow rate $\frac{dm}{dt}$ into or out of the pipeline equates to zero. Applying the chain rule for the ideal gas law, the mass conservation equation in rate form becomes:

$$\frac{dm}{dt} = \frac{pV}{RT}\left[\left(\frac{1}{p}\right)\frac{dp}{dt} + \left(\frac{1}{V}\right)\frac{dV}{dt} - \left(\frac{1}{T}\right)\frac{dT}{dt}\right] = 0 \quad \textbf{(1)}$$

In Eq. (1), m, p, V and T are the mass, pressure, volume and temperature of air in the LPACP, respectively, all expressed in SI units. The derivatives thereof thus represent the change of the given air parameter over a time increment dt. The rate of change of internal energy of the system could further be given via the energy balance equation:

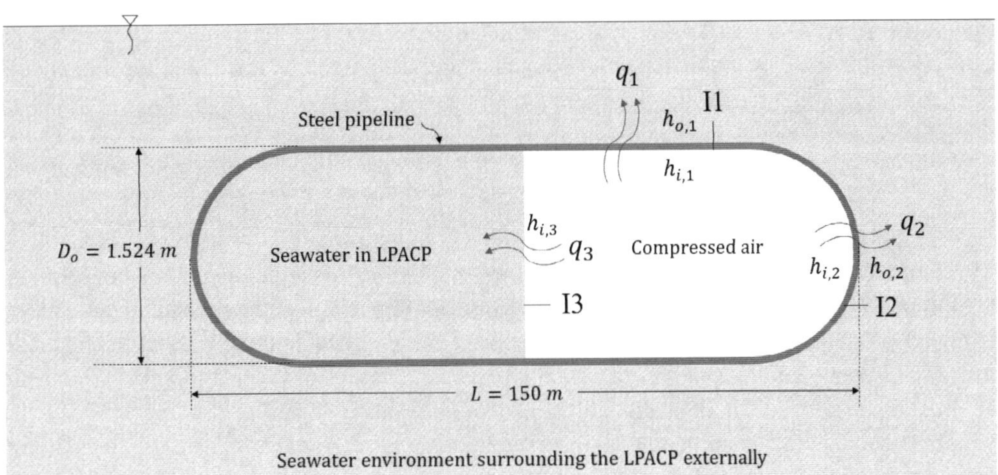

Figure 1. Geometry, dimensions and thermal interfaces for a subsea LPACP. Diagram is not to scale.

 doi:10.1088/1742-6596/2929/1/012001

$$mc_v \frac{dT}{dt} = \frac{dQ}{dt} - p\frac{dV}{dt} + \frac{dW_{hyd}}{dt} \tag{2}$$

In Eq. (2), c_v is the specific heat at constant volume, Q is the heat transfer from the body of air and W_{hyd} is the work done by hydrostatic pressure due to a subsea LPACP. Indeed, the last two terms on the right-hand side of Eq. (2) symbolise the total work input to the system. The term $\frac{dV}{dt}$ equates to the prescribed seawater volume flow rate into the pipeline derived from the affixed power input P and the pressure head across the hydraulic pump, Δp_{pump}:

$$P = \Delta p_{pump} \frac{dV}{dt} \tag{3}$$

Solving for the heat transfer rate $\frac{dQ}{dt}$ as summarised in Section 2.2, Eqs. (1) and (2) could be solved as a pair of simultaneous equations to determine the unknown rate of air pressure and temperature change at any instance of compression.

2.2 Heat transfer analysis

The thermal model considered heat transfer via interfaces *I1* to *I3* (refer to Figure 1) attributable to the energy storage medium (i.e., air). The heat transfer interfaces in contact with the internal seawater were not considered since rigorous testing had confirmed that the change in temperature of the aforesaid fluid is minimal. Besides, the seawater does not contribute to any storage of energy. The total heat transfer rate from the compressed air $\frac{dQ}{dt}$ equates to the summation of the heat transfer rate across the cylindrical part (*I1*), one hemi-spherical end (*I2*) and the air-seawater interface (*I3*). In mathematical form:

$$\frac{dQ}{dt} = q_1 + q_2 + q_3 \tag{4}$$

where q_1, q_2 and q_3 are expressed in Watts. Every term on the right-hand side of Eq. (4) was determined by solving the thermal circuit developed for the LPACP. In the case of q_1 and q_2, the parameters accounted for (i) the convective heat transfer between the air and the internal steel walls, (ii) the conductive heat transfer across the steel wall and (iii) the convective heat transfer between the external steel surface and surrounding seawater. The HTC for every interface, both externally and internally (as applicable), was calculated separately. Indeed, a total of five HTCs (i.e., $h_{o,1}$, $h_{i,1}$, $h_{o,2}$, $h_{i,2}$, $h_{i,3}$) was considered, as indicated on Figure 1. In general, all HTCs could be found in terms of the Nusselt (Nu) number:

$$h = Nu\frac{k}{D} \tag{5}$$

In Eq. (5), k is the thermal conductivity of the fluid in question and D is the pipeline diameter. At the time of writing, correlations for the *Nu* number specifically for subsea LPACP of large scales were found to be lacking in open literature. Consequently, the numerical model was based on existing empirical formulas that, after an extensive review, were deemed to best fit the intended application. Table 1 summarises all the correlations and/values of HTCs adopted at the different heat transfer interfaces, for the different modes of convection. From preliminary testing, it was observed that mixed convection effects occurred for very short periods of time (i.e., less than 10 s) during one stroke. Consequently, for a simplified and a conservative modelling approach, the short-lived mixed convection instance per stroke was assumed to be dominated by free convection phenomena.

Table 1. List of correlations implemented at the different LPACP interfaces

Location	Type of convection	HTC	Correlation/Fixed value	Source
External	Natural	$h_{o,1}$	$Nu = \left\{ 0.60 + \frac{0.387 Ra^{\frac{1}{6}}}{\left[1 + \left(\frac{0.559}{Pr}\right)^{\frac{9}{16}}\right]^{\frac{8}{27}}} \right\}^2$	Churchill-Chu [15]
External	Forced	$h_{o,1}$	$Nu = 0.30 + \left\{ \frac{0.62 Re^{\frac{1}{2}} Pr^{\frac{1}{3}}}{\left[1 + \left(\frac{0.4}{Pr}\right)^{\frac{2}{3}}\right]^{\frac{1}{4}}} \right\} \left[1 + \left(\frac{Re}{282{,}000}\right)^{\frac{5}{8}}\right]^{\frac{4}{5}}$	Churchill-Bernstein [16]
External	Natural	$h_{o,2}$	$Nu = 0.533 Ra^{0.25}$	Lewandoski et al. [17]
External	Forced	$h_{o,2}$	$Nu = 2 + \left(0.4 Re^{\frac{1}{2}} + 0.06 Re^{\frac{2}{3}}\right) Pr^{0.4} \left(\frac{\mu}{\mu_w}\right)^{\frac{1}{4}}$	Whitaker [18]
Internal	Natural	$h_{i,1}$	$Nu = 1.15 Ra^{0.22}$	Ludovisi-Garza [19]
Internal	Forced	$h_{i,1}$	$Nu = \frac{\left(\frac{f}{8}\right)(Re - 1000) Pr}{1 + 12.7 \left(\frac{f}{8}\right)^{\frac{1}{2}} \left(Pr^{\frac{2}{3}} - 1\right)}$	Gnielinski [20]
Internal	Natural	$h_{i,2}$	$Nu = 0.2357 Ra^{0.242}$	Shiina et al. [21]
Internal	Forced	$h_{i,2}$	100 W/m² K	[22]
Internal	Natural/Forced	$h_{i,3}$	200 W/m² K	[6]

For the computation of the *Nu* number attributable to $h_{o,1}$ widely-used formulas for cylindrical bodies in cross-flow were implemented. Indeed, the Churchill and Chu [15] correlation and Churchill and Bernstein [16] relationship were adopted for natural and forced convection, respectively, across *I1*. The formula derived by Lewandoski et al. [17] for hemi-spherical bodies was chosen to compute the *Nu* number related to $h_{o,2}$ under natural convection conditions. On the other hand, to the best of the authors' knowledge, definitions of the *Nu* number around hemi-spherical caps under forced convection effects are currently lacking. Consequently, the next best approach was to implement the formula proposed by Whitaker [18] which simulates forced convective flow of fluids over spheres.

 doi:10.1088/1742-6596/2929/1/012001

The derivation of the internal HTCs was somehow more cumbersome to obtain due to insufficient published work. The determination of $h_{i,1}$ under natural convection states was based on the Nu number definition proposed by Ludovisi and Garza [19]. On the other hand, the Gnielinski [20] correlation was utilised for the computation of $h_{i,1}$ under forced convection phenomena. The formulation by Shiina et al. [21] was assumed to best fit the internal free convection effects at the hemi-spherical ends. At the time of writing, no existing correlations explicit to internal forced convection in neither hemi-spherical nor spherical bodies were found in the public domain. Therefore, a constant value of 100 W/m^2 K was prescribed. The quoted figure was chosen based on typical HTC values for forced convection of air as reported in literature [22]. Similarly, a substantial dearth of information was noted in relation to formulas or general magnitudes related to the HTC at the air-seawater interface (i.e., $h_{\text{i},3}$). Li et al. [6] estimated a constant value of 196 W/m^2 K, specifically for the air-water interface inside a LPAC cylinder. The rounded-up value of 200 W/m^2 K was thus implemented for $h_{\text{i},3}$ in the present model.

2.3 Key performance indicators (KPIs)

To adequately assess the thermal behaviour of the LPACP at high pressure ratios, three key performance indicators (KPIs) were used. First, is the polytropic index of compression n evaluated via Eq. (6) and which can vary between a minimum value of unity for an isothermal process, and a maximum value of 1.4 (for air) for an adiabatic compression. In Eq. (6), r_p and r_v are the pressure (of air) and volume (of air) ratios of operation, respectively. Moreover, $p_{final} > p_{initial}$ and $V_{final} < V_{initial}$. Consequently, the closer n is to a value of 1, the more likely the system is to operate close to ideal, isothermal conditions.

$$n = \frac{\log\left(\frac{p_{final}}{p_{initial}}\right)}{\log\left(\frac{V_{initial}}{V_{final}}\right)} = \frac{\log r_p}{\log r_v} \tag{6}$$

In conjunction with the value of n, the maximum air temperature $T_{a,max}$ reached at the end of the LP compression stroke is indicative of the system thermodynamic performance. The closer $T_{a,max}$ remains to the intial value (equivalent to the seawater temperature), the more isothermal the process is. Another KPI is the thermodynamic efficiency η_{th}, which is defined as the ratio of the ideal to the real work done:

$$\eta_{th} = \frac{W_{iso}}{W_{real}} = \frac{p_{final}V_{final}\ln(r_p) - p_{hyd}V_{final}(r_p - 1)}{\int_{V_{initial}}^{V_{final}} pdV - p_{hyd}V_{final}(r_p - 1)} \tag{7}$$

In Eq. (7), the term p_{hyd} is the hydrostatic pressure that must be considered for a subsea LPACP.

Under real conditions, the work required for compression is always greater than the ideal work since additional work needs to be done due to the losses incurred. Subsequently, the less real work is done, and the closer W_{real} is to W_{iso}, the higher the value of η_{th}.

3. Results

Following verification procedures and validation against the work of Chen et al. [7] for one compression cycle of the LPACP, a parametric study was carried out. The geometry and dimensions of the LPACP were maintained constant throughout the investigation, as indicated on Figure 1. The LPACP was considered to have an external diameter D_o of 1.524 m and a length of 150 m, thus having an L/D_o ratio of 98. Other default test parameters retained fixed included the

Journal of Physics: Conference Series **2929** (2024) 012001 doi:10.1088/1742-6596/2929/1/012001

initial air pressure and temperature of 1 bar and 293.15 K, respectively. The air temperature was also initially assumed to be in thermal equilibrium with the steel pipeline and external seawater temperatures. The HTCs were defined as indicated in Table 1 of Section 2.2. The pipeline was considered to be composed of steel grade X70, with a thermal conductivity of 64 W/m K and a specific heat capacity of 480 J/kg K [23]. The test sea depth was fixed at 10.5 m with the external seawater flow represented by an orbital wave particle velocity of 0.61 m/s [24] representing rough sea states in accordance with [25]. Contrastingly, the input power P, and maximum operating pressure p_{final} (and thus pressure ratio r_p) were parametrised as indicated in Sections 3.1 to 3.3 to investigate the influence on the KPIs, and thus, on the thermal performance of the LPACP.

3.1 Sensitivity of KPIs to power density (PD) and pressure ratio (r_p)

For the sensitivity analysis, the input hydraulic power P via the LP, calculated via Eq. (3), was altered from a maximum of 420 kW down to 6.56 kW, each time halving the previous value. Therefore, with a fixed LPACP volumetric capacity of 273.70 m^3, the power density (PD) was varied between 1.77 and 0.03 kW/m^3. In turn, for every PD (or power input), p_{final} was set to 200, 150, 100 and 50 bar. Consequently, the pressure ratios r_p tested were 200, 150, 100 and 50, respectively, given that the initial pressure $p_{initial}$ was always set to the atmospheric value of 1 bar absolute.

Figure 2 presents the non-linear relationship between the polytropic index, n and PD, for the different values of p_{final}. The plot firstly reveals that n is generally insensitive to the magnitude of p_{final} as the curves for the 200, 150, 100 and 50 bar are almost perfectly superimposed. However, Eq. (6) suggests that n is highly dependent on r_p, and thus p_{final}, as well as r_v.and thus V_{final}. Essentially, with a change in p_{final} and thus, r_p, the corresponding final volume V_{final} upon end of compression changes accordingly as shown in Figure 3. Thus, the value of r_v adjusts in a manner that with a change in r_p, n remains unaltered. Analysing the trend in Figure 2, it is also evident that the air compression in the LPACP approaches isothermal conditions with lower PDs. The latter statement is well in line with what has been recorded in literature [6-9] for small-scale LPACs operating in open atmosphere and at lower pressure ratios. Lower PDs have been attributed to lengthier compression periods, allowing sufficient heat dissipation to and from the LPAC to maintain minimal temperature changes. Indeed, the plot in Figure 4 demonstrates the exponential rise in compression time with reduced PDs (or power inputs).

Figure 5 further confirms the results presented in Figure 2, where a decreasing PD is observed to maintain the maximum air temperature $T_{a,max}$ closer to the initial temperature (and surrounding seawater temperature) which during the parametric analysis was kept at 293.15 K. Further evidence is provided in Figure 6 where the highest PDs and pressures lead to the lowest thermodynamic efficiencies, η_{th}. For instance, with a PD of 1.77 kW/m^3, the air compression from 1 to 200 bar is predicted to reach an efficiency η_{th} of 0.75. Reducing the PD down to 0.03 kW/m^3, the system can attain an exceptional efficiency of 0.97 for the same pressure range. Considering the charging cycle from 1 to 50 bar, at a maximum PD of 1.77 kW/m^3, the system reaches an efficiency of 0.87. The latter is 16% higher than the efficiency of 0.75 when pressurising from 1 to 200 bar at the same PD (i.e., using the same power input of 420 kW). Interestingly, Figure 6 also reveals that, despite situating the LPACP subsea, and despite utilising a $L/D_o \gg 1$, η_{th} can still be significantly lower than 0.90 when r_p > 50. In fact, considering all test r_p, the maximum allowable PD to reach η_{th} > 0.90, equates to 0.22 kW/m^3. The latter translates into 25 to 72 minutes of compression time, depending on the desirable p_{final}. Furthermore, based on the work done per compression stroke, the energy transferred per stroke was found to vary between 17 and 35 kWh/stroke as can be observed on Figure 7. Comparing the outcomes in Figures 6 and 7, it may be realised that the higher energy requirement (and hence work done), the lower the value of η_{th}, since a higher amount of energy input is being converted into heat.

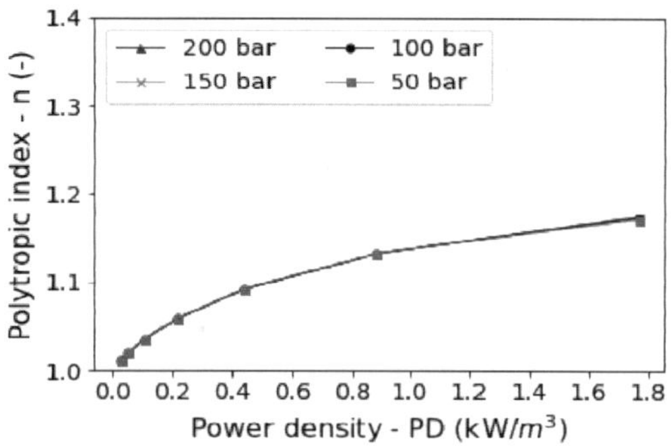

Figure 2. Polytropic index variation with PD, for different peak pressures

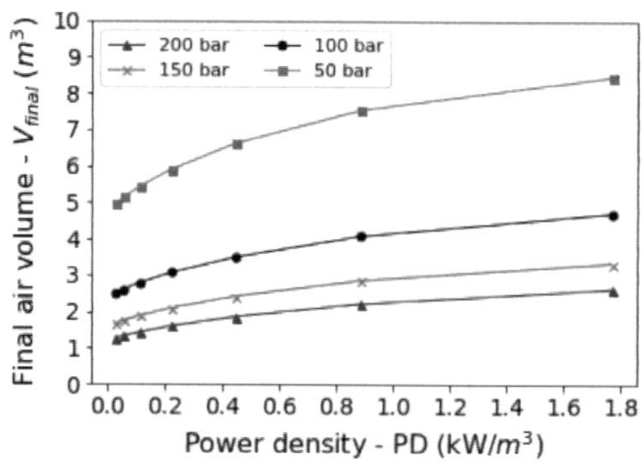

Figure 3. Final air volume variation with PD, for different peak pressures

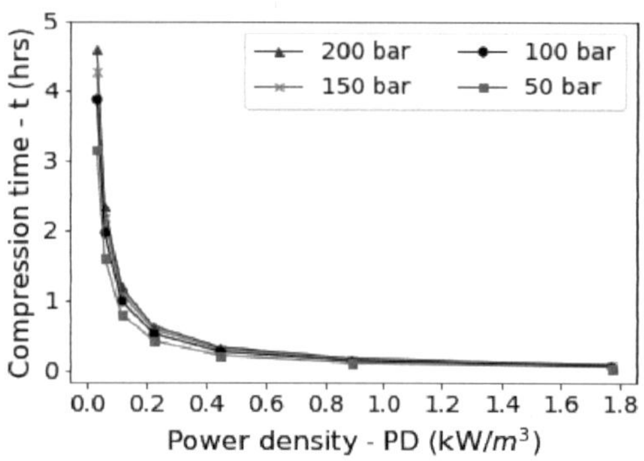

Figure 4. Compression time variation with PD, for different peak pressures

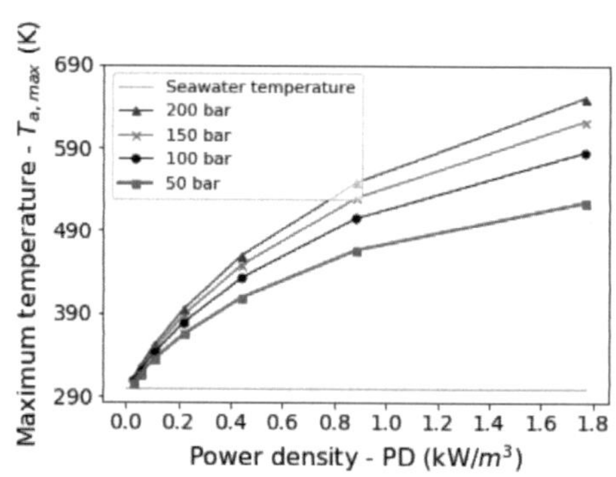

Figure 5. Maximum air temperature variation with PD, for different peak pressures

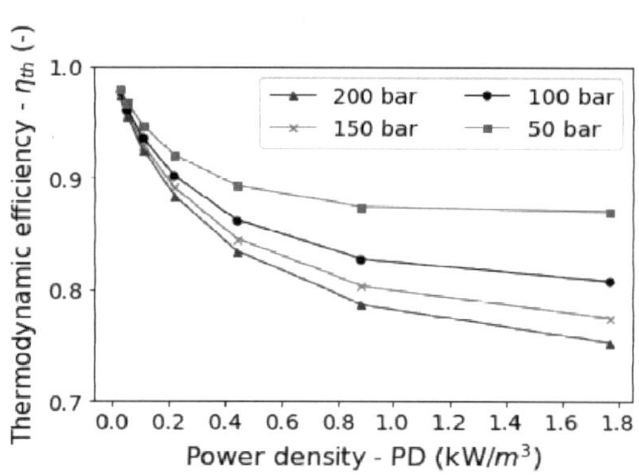

Figure 6. Thermodynamic efficiency variation with PD, for different peak pressures

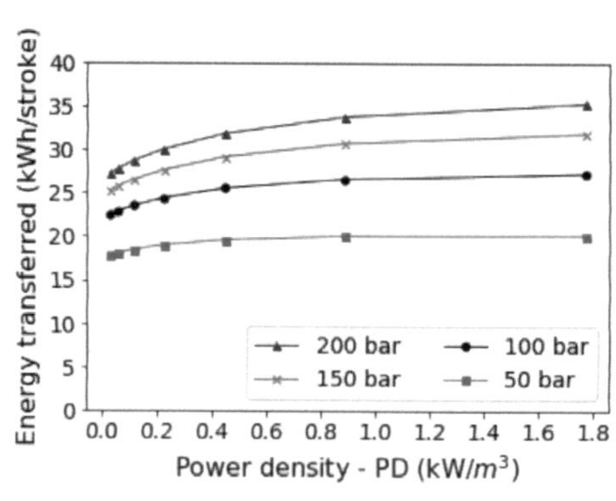

Figure 7. Variation of energy transferred per stroke with PD, for different peak pressures

3.2 Understanding the corresponding change in external HTCs with PD and r_p

This section investigates the sensitivity of the external HTCs to the PD and r_p (or p_{final}). The values for the PD, and thus, input hydraulic power P, were varied in an identical manner to the magnitudes listed in Section 3.1.

Firstly, Figures 8 and 9 display the variation of $h_{o,1}$ and $h_{o,2}$ with compression time, respectively. The outcomes for the 200 bar peak pressure are presented as an example. On the other hand, Figures 10 and 11 portray the variation of the peak value of external HTCs with PD, for a 200 and a 50 bar peak pressure. The magnitudes of HTCs on the aforesaid figures indicate that, externally, the seawater conditions prescribed as highlighted in Section 3, were representative of forced convection due to the seawater velocity, where the fluid is hardly ever quiescent. Therefore, the values of $h_{o,1}$ and $h_{o,2}$ shown on Figures 8 to 11 were estimated from the Churchill-Bernstein [16] and Whitaker [18] formulas, respectively (refer to Table 1).

3.3 Understanding the corresponding change in internal HTCs with PD and r_p

For the same test conditions of Sections 3.1 and 3.2, Figures 12 and 13 exhibit the change in $h_{i,1}$ and $h_{i,2}$ with compression time, respectively, for the 200 bar peak pressure scenario. The maximum magnitudes of internal HTCs with different PDs can then be observed on Figures 14 and 15. The internal heat transfer mode from the compressed air to the surroundings was found to be dominated by the buoyancy effects which start to take over the pump-driven forces for reducing powers/PDs. Thus, the values of $h_{i,1}$ and $h_{i,2}$ as per Figures 12 to 15 were derived from the Ludovisi-Garza [19] and Shiina et al. [21] formulas, for internal natural convection, respectively (refer to Table 1). It is however crucial to emphasise that, the natural convection mode is subject to the geometry of the LPACP under investigation, having an L/D_o ratio of 98.

3.4 General comments for internal and external HTCs variation with PD and r_p

In general, Figures 8 to 15 also show how the peak HTC values are larger in magnitude for higher pressures and PDs. In Section 3.1, it has been shown that a reduction in PD translates into longer charging durations, and consequently, higher n_{th}. Additionally, higher values for the HTC are typically attributed to values of n_{th} approaching unity. However, Figures 8 to 15, indicate smaller HTC peak values with smaller PDs, which has a paradoxical effect on n_{th} (i.e., smaller HTC values decrease n_{th} but smaller PDs improve n_{th}). Yet, Figure 6 demonstrates that the overall outcome is an improvement in n_{th}. Thus, it is the power/charging duration which is having the leading impact on n_{th}. In general, it can be concluded that, the increased thermodynamic efficiency with a reduction in PD highlighted in Section 3.1, is governed by longer compression durations rather than improved HTC values. The magnitude of PD (or input power) does influence the value of HTCs at the different interfaces, yet the change in HTC is not sufficient to affect n_{th} considerably. A further remark is that, for a given case study, the HTCs at the hemi-spherical end *I2*, are smaller than the corresponding HTCs at the cylindrical surfaces *I1* (i.e., $h_{o,1} > h_{o,2}$ and $h_{i,1} > h_{i,2}$). The differences in the magnitude for the HTC result from the different correlations being used (refer to Table 1) used. The correlations from [18, 21] utilised for *I2* yield smaller values for the Nu number, $h_{o,2}$ and $h_{i,2}$, than the equivalent values derived from the correlations of [16, 19] for *I1*. Recalling the elucidation in Section 2.2, the investigation of $h_{i,3}$ is not being presented as the parameter was prescribed a fixed value of 200 W/m^2 K (refer to Table 1) due to the inadequacy of information and empirical correlations available.

Journal of Physics: Conference Series **2929** (2024) 012001 doi:10.1088/1742-6596/2929/1/012001

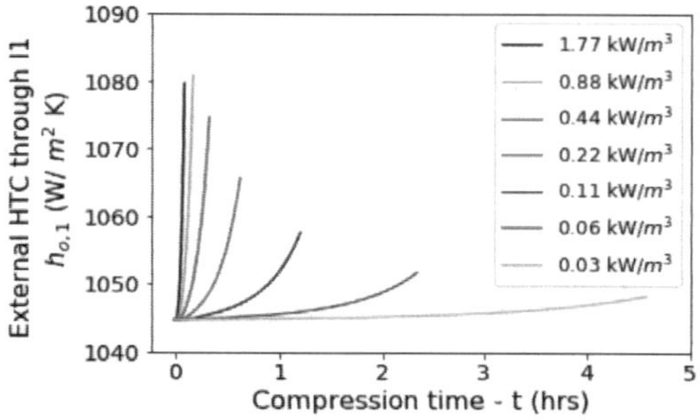

Figure 8. Variation of $h_{o,1}$ with compression time for a 200 bar peak pressure

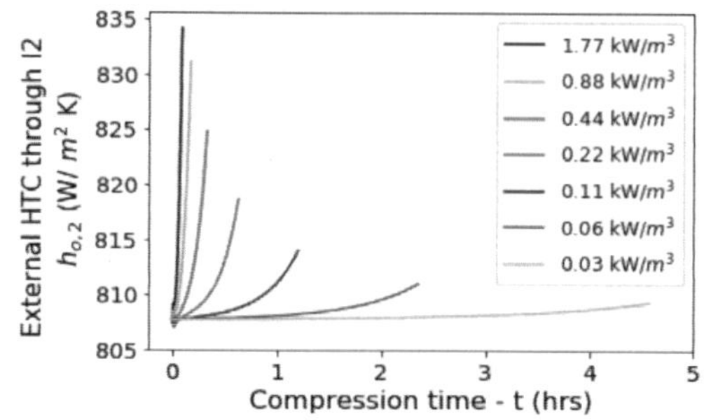

Figure 9. Variation of $h_{o,2}$ with compression time for a 200 bar peak pressure

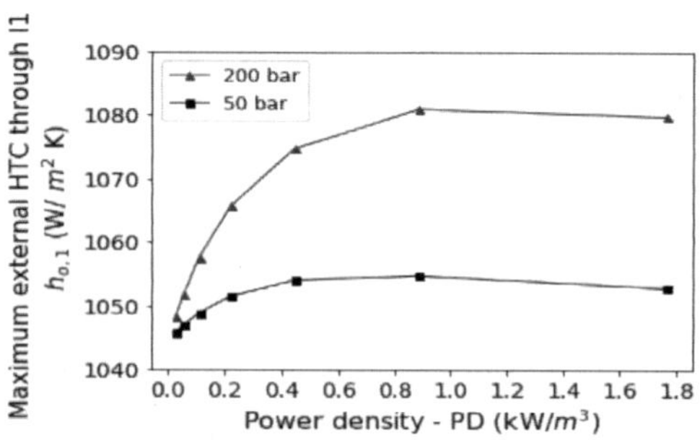

Figure 10. Variation of maximum $h_{o,1}$ value with PD, for 50 and 200 bar peak pressures

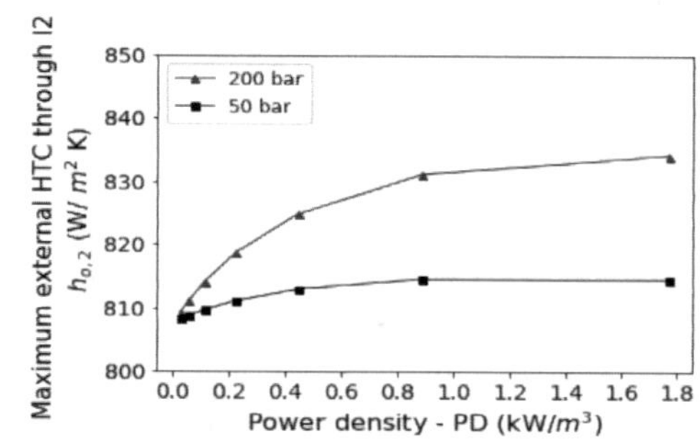

Figure 11. Variation of maximum $h_{o,2}$ value with PD, for 50 and 200 bar peak pressures

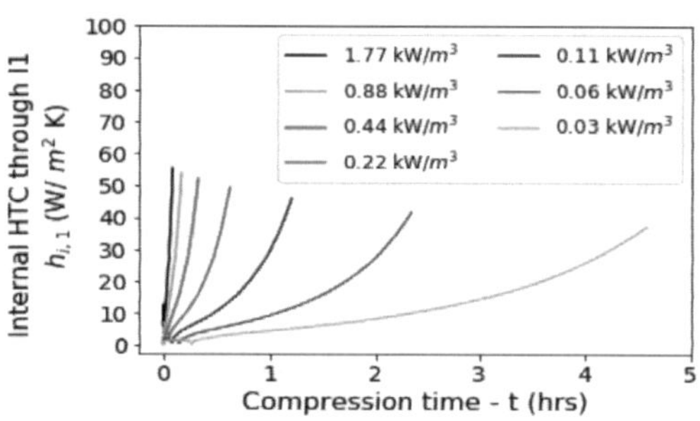

Figure 12. Variation of $h_{i,1}$ with compression time for a 200 bar peak pressure

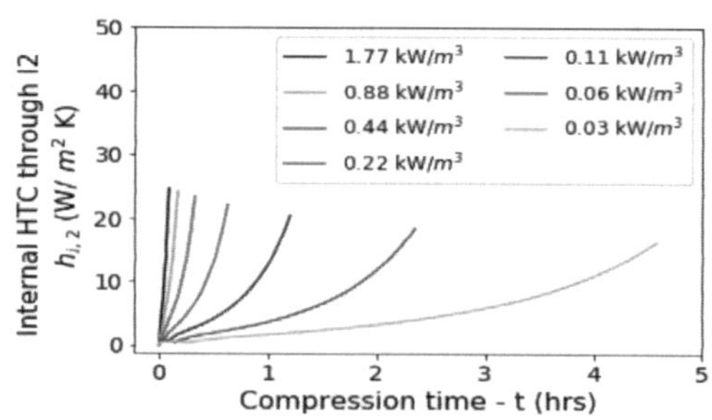

Figure 13. Variation of $h_{i,2}$ with compression time for a 200 bar peak pressure

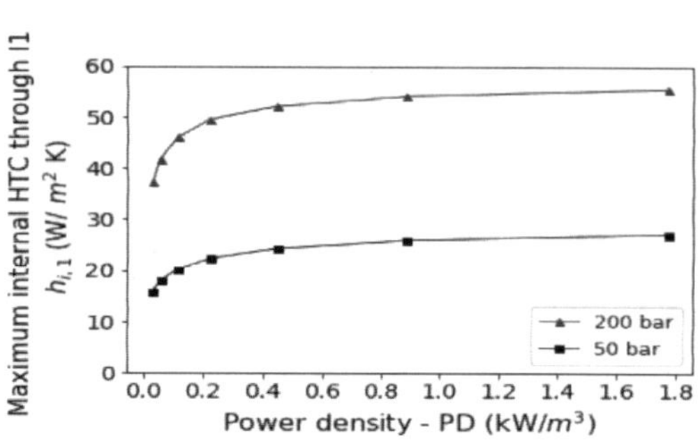

Figure 14. Variation of maximum $h_{i,1}$ value with PD, for 50 and 200 bar peak pressures

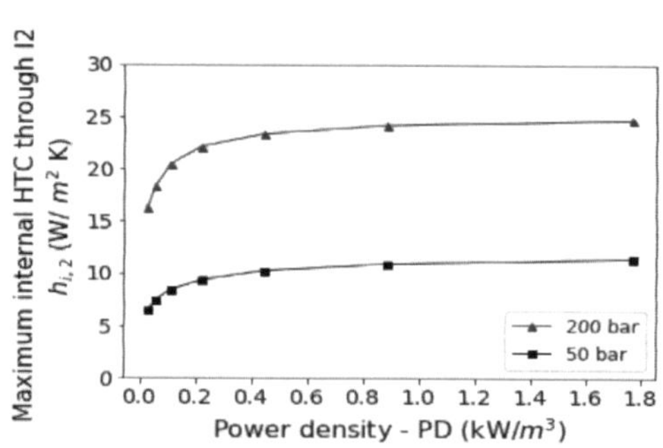

Figure 15. Variation of maximum $h_{i,2}$ value with PD, for 50 and 200 bar peak pressures

4. Conclusion

Liquid piston air compression is gaining research interest due to its application in emerging OI-CAES technologies. Yet, the difficulty in finding appropriate heat transfer correlations for the convective processes on the inner surface of the LPAC is a major knowledge gap in literature for reliable modelling of the thermal performance of such systems. The study applied a transient thermal model for a subsea LPACP operating with high pressure ratios, based on existing correlations that best approximate the subject application. The main conclusions from the numerical investigation include:

i. operating at higher r_p values results in lower thermodynamic efficiencies as the air temperature rise in the LPACP occurs faster for a given power input/PD. Indeed, for relatively high powers/PDs, thermodynamic efficiencies lower than 0.80 were estimated when compressing air subsea from 1 to 200 bar;
ii. increasing the input power/PD will significantly effect the compression duration but is not sufficient to cause a change in the order of magnitude of the HTCs. Consequently, increased powers/PDs result in lower thermodynamic efficiencies due to insufficient time for heat transfer;
iii. despite situating the LPACP subsea, and despite utilising a $L/D_o \gg 1$, η_{th} can still be significantly lower than 0.90 when $r_p > 50$. In fact, considering all test pressure ratios, the maximum allowable PD to reach $\eta_{th} > 0.90$, equates to 0.22 kW/m^3. The latter translates into 25 to 72 minutes of compression time, depending on the desirable p_{final};
iv. the values of HTC at the hemi-spherical end (*I2*) were observed to be slightly less than the HTC values at the cylindrical interface (*I1*);
v. externally, a subsea LPACP is more likely to experience forced convection due to the seawater conditions in which the fluid is rarely ever quescient;
vi. internally, the heat transfer from the compressed air to the surroundings during compression is dominated by natural convection for a LPACP having a L/D_o ratio of 98.

More importantly, the study has highlighted a major knowledge gap in terms of correlations available to explicitly determine the Nu number (and thus HTCs) inside a LPAC. Furthermore, the determination of HTC values at the air-water interface are also well lacking. It is thus being suggested that further research, including CFD and experimental testing on large-scale LPACPs operating at high pressure ratios be conducted to provide better predictions on the expected performance of LPACP and/or LPACs in general. The future work being suggested can provide additional information on aspects which have been simplified during the present work. For example, CFD analysis would be able to capture the thermal gradients of the fluid(s) across the LPACP. In addition, the inlets and outlets to and from the LPACP can be further incorporated to investigate any implications on the thermal performance of the system.

Acknowledgements

The research presented in this paper forms part of project SAICOPES (Subsea Air Isothermal Compression using Offshore Pipelines in Long Duration Energy Storage Applications) funded by the Malta Council for Science and Technology (MCST) under the Research Excellence Programme. Project Reference: REP-2023-004.

References

[1] IEA, "Energy storage," IEA, https://www.iea.org/energy-system/electricity/grid-scale-storage (accessed Apr. 5, 2024).
[2] Sadaat M and Li P Y, "Modeling and Control of a Novel Compressed Air Energy Storage System for Offshore Wind Turbine," *in the 2012 American Control Conference*, Montreal, Canada, 2012.
[3] Odukomaiya A *et al* "Experimental and Analytical Evaluation of a Hydro-pneumatic Compressed-Air Ground-Level Integrated Diverse Energy Storage (GLIDES) System," *Applied Energy*, vol. 221, pp. 75-85, 2018.
[4] "Enabling large-scale offshore energy storage," FLASC – Renewable Energy Storage, https://offshoreenergystorage.com/ (accessed Apr. 11, 2024).
[5] Budt M *et al* "A Review on Compressed Air Energy Storage: Basic Principles, Past Milestones and Recent Developments," *Applied Energy*, vol. 170, pp.250-268, 2016.
[6] Li C *et al* "Experimental and Thermodynamic Investigation on Isothermal Performance of Large-scaled Liquid Piston," *Energy*, vol. 249, pp. 123731, 2022.
[7] Chen H *et al*, "Thermodynamic Analysis of an Open Type Isothermal Compressed Air Energy Storage System based on Hydraulic pump/turbine and Spray Cooling," *Energy Conversion and Management*, vol. 204, pp. 112293, 2020.
[8] He X *et al*, "Thermodynamic Investigation of Variable-Speed Compression Unit in Near-Isothermal Compressed Air Energy Storage," *Energy Storage*, vol. 5, no. 8, 2023.
[9] Van de Ven J D and Li P Y, "Liquid Piston Gas Compression," *Applied Energy*, vol. 86, pp. 2183-2191, 2009.
[10] "Welcome to Python.org," Python.org, https://www.python.org/ (accessed Apr. 25, 2024).
[11] Cutajar C et al., "A Software Tool for the Design and Operational Analysis of Pressure Vessels used in Offshore Hydro-pneumatic Energy Storage," Journal of Energy Storage, vol. 40, pp. 102750, 2021.
[12] American Petroleum Institute, Specification for Line Pipe, 43rd Edition, API Specification 5L, 2004.
[13] European Norm, Unfired Pressure Vessels – Part 2: Materials, European Standard (2014). SM EN 13445-2:2014.
[14] Directive, 2014/68/EU of the European Parliament and of the Council of 15 May 2014 on the harmonisation of the laws of the Member States relating to the making available on the market of pressure equipment, Official Journal of the European Union, Brussels, Belgium (2014).[15] Churchill S W and Chu H H S, "Correlation Equations for Laminar and Turbulent Free Convection from a Horizontal Cylinder," *International Journal of Heat and Mass Transfer*, vol. 18, no. 9, pp.1049–1053, 1975.
[16] Churchill S W and Bernstein M, "A Correlating Equation for Forced Convection from Gases and Liquids to a Circular Cylinder in Crossflow," *Journal of Heat Transfer*, vol. 99, no. 2, pp.300-306, 1977.
[17] Lewandoski W M *et al*, "Theoretical and Experimental Study of Natural Convection Heat Transfer from Isothermal hemispheres," *International Journal of Heat and Mass Transfer*, vol. 40, no. 1, pp: 101-109, 1997.
[18] Whitaker S, "Forced Convection Heat Transfer Correlation for Flow in Pipes, Past Flat Plates, Single Cylinders, Single Spheres, and for Flow in Packed Beds and Tube Bundles," *AIChE Journal*, vol. 18, pp. 361, 1972.
[19] Ludovisi D and Garza I A, "Natural Convection Heat Transfer in Horizontal Cylindrical Cavities: A Computational Fluid Dynamics (CFD) Investigation," *in Proceedings of the ASME 2013 Power Conference*, Massachusetts, United States, 2013.
[20] Gnielinski V, "New Equations for Heat and Mass Transfer in Turbulent Pipe and Channel Flow," *International Chemical Engineering*, vol. 16, no. 2, pp. 359-68, 1976.
[21] Shiina Y *et al*, "Natural Convection Heat Transfer in Hemisphere," *Journal of Nuclear Science and Technology*, vol. 25, no. 3, pp. 254-262, 1988.
[22] Incropera F P *et al*, *Fundamentals of Heat and Mass Transfer*, 6th ed., John Wiley and Sons, Hoboken, United States, 2007.
[23] Li L *et al*, "Generation Mechanism of Lack of Fusion in X70 Steel Welded Joint by Fully Automatic Welding under Steep Slope Conditions Based on Numerical Simulation of Flow Field," *The International Journal of Advanced Manufacturing Technology*, vol. 126, pp. 4055-4072, 2023.
[24] Solving Complex, Real-World Problems. BMT. Available online: https://www.bmt.org/ (accessed on 26 September 2024).
[25] *Manual on Codes, International Codes*, 2019 ed.; World Meteorological Organization, Volume I.1, Geneva, Switzerland, 2019.

CFD Analysis of a Subsea Liquid Piston Accumulator for Energy Storage Applications

Luke Aquilina[1*], Tonio Sant[1] and Charise Cutajar[1]

[1] Department of Mechanical Engineering, Faculty of Engineering, University of Malta, Malta

*E-mail: luke.aquilina@um.edu.mt

Abstract. Hydro-pneumatic energy storage uses liquid pistons and hydraulic machinery to store energy by compressing air. The technology offers significant potential for co-location with offshore wind farms, where the sea water may be used to augment the heat exchange between the air undergoing compression or expansion and the surrounding environment, hence improving the thermal efficiency by operating at isothermal conditions. A major challenge for the numerical simulations to assess the transient thermal behaviour of such a system is the lack of appropriate models for the convective heat transfer coefficients. This paper addresses this knowledge gap through the use of CFD by simulating a subsea accumulator housing a liquid piston compressing the air from atmospheric conditions up to a pressure of 100 Bar. The adopted numerical approach examines the heat exchange between the air undergoing compression and the surrounding sea water, with the internal and external heat transfer coefficients being derived concurrently. It is noted that the internal heat transfer coefficient is significantly dependent on the state of charge of the accumulator. Yet, quasi-isothermal conditions were predicted by the CFD model for the simulated conditions. The paper provides insight about the complexity of the heat transfer phenomena related to subsea liquid pistons and the need to derive correlations specifically for this application.

Keywords. Liquid Piston Compression, Computational Fluid Dynamics, Hydro-Pneumatic Energy Storage, Convective Heat Transfer

List of Abbreviations.

CFD	Computational Fluid Dynamics
ESS	Energy Storage System
HPES	Hydro Pneumatic Energy Storage
HTC	Heat Transfer Coefficient
RES	Renewable Energy Sources
VOF	Volume of Fluid

List of Symbols.

D	Accumulator internal diameter, m
h_{ext}	External convective HTC, W/m²K
h_{int}	Internal convective HTC, W/m²K
h_{int0}, h_{intB}	Internal convective HTC calculated using T_0 and T_{BULK} respectively, W/m²K
L	Accumulator cylindrical length, m
$\dot{m}$	Mass flow, kg/s
P	Inlet hydraulic power, W
p	System pressure, Pa
p_{hyd}	Hydrostatic pressure, Pa
Q	Flow rate, m^3/s
q_{int}, q_{ext}	Internal and External wall heat fluxes respectively, W/m2
R	Accumulator internal radius, m
t	System time, s
T_0, T_{50}, T_{75}	Accumulator internal temperatures taken at 0, 50 and 75% R, K
T_a, T_w	Air and water temperatures within accumulator respectively, K
T_{wi}, T_{we}	Accumulator internal and external wall surface temperature, K
T_{SW}	Surrounding water temperature, K
V_{wr}	Accumulator water volume ratio
v	System air volume, m^3

1. Introduction

Studies into the integration of Renewable Energy Sources (RES) such as wind and solar power have however indicated that the intermittent nature of such energy sources would pose significant challenges to the national power grids, particularly at high levels of penetration[1, 2, 3]. Energy Storage Systems (ESS) will become essential to mitigate issues related to the integration of RES into national grids [4, 5, 6]. ESSs work by primarily storing excess energy during periods of low demand, with this energy then being used during periods of high demand, thus acting as a buffer to allow RES to match the demand from the grid [5]. There are many forms of ESSs that are currently being developed to address this pertinent issue [4, 5].

Hydro-Pneumatic Energy Storage (HPES) is one example of mechanical ESS technology where energy is stored and released by compressing and expanding air via a liquid piston [7, 8, 9]. The system offers potential to be deployed offshore, using a pre-charged accumulator placed below the sea surface [10]. Energy transfer to and from the system takes place using hydraulic machinery. During the charge cycle, electrical power is converted into hydraulic power through the use of a hydraulic pump that injects water into the accumulator to form a liquid piston that compresses the stored air. During the discharge cycle, the stored air is allowed to expand by allowing the pressurised water in the accumulator to flow through a hydraulic turbine to generate power and supply electricity to the grid. Previous studies involving both numerical modelling and physical experiments have demonstrated that when such systems are deployed in open sea water, quasi-isothermal conditions can be achieved during both charging and discharging, reaching round-trip thermal efficiencies as high as 93% [10]. In this case, the term thermal efficiency was used instead of an overall system efficiency as the former refers only to the process of compression and decompression of the trapped air as analysed from a thermodynamic aspect alone, whilst excluding losses pertaining to hydraulic pump, turbine and other ancillary components.

Various studies have been conducted on liquid pistons using both Computational Fluid Dynamics (CFD) and physical experiments to better understand the complex heat transfer interaction between the liquid piston and the air within the accumulator. Gouda et al. [11] and Khaljani et al. [12] carried out physical experiments on liquid pistons and compared these to CFD experiments. In the case of Gouda et al., the physical experiments used a simple mechanical piston (not a liquid piston). In spite of this however, results compared well with CFD simulations of a similar system modelling a liquid piston. It was noted that for such a liquid piston to work efficiently, the process needs to be slow enough to allow for adequate heat transfer between the accumulator and its surroundings. As for Khaljani et al., the experiments used an accumulator set up with integrated aluminium plates to help improve the heat transfer between the fluids inside the accumulator and the accumulator walls. In this case, CFD results were also in agreement with physical experiments. From these experiments it was noted once again that improving the heat transfer between the air inside the accumulator and its surroundings, in this case through the use of the aluminium plates, increases the overall efficiency of the system.

Mutlu et al. [13] carried out CFD experiments on accumulators undergoing compression using liquid pistons. For their experiments, Mutlu et al. focussed on the effect of different accumulator geometries, as well as the water injection flow rate into the vessel. From such experiments, it was observed that accumulators with larger aspect ratios offer better operational efficiencies, particularly if combined with slower water injection flow rates. Similarly, Zhang et al. [14] observed from CFD experiments that smaller accumulators achieve improved thermal efficiencies. An interesting aspect of the experiments conducted by Zhang et al. was the inclusion of a porous metal foam within the vessel to help increase the heat transfer between the fluids inside the vessel and the surroundings. Such an improvement was also experimented upon by Wieberdink et al. [15] where a physical model of a combined compressor and expander setup using a liquid piston was set up with a stainless steel mesh to improve the heat transfer within the liquid piston. Results have indicated that such an insert could drastically improve the efficiency of the system, with the ideal location of the mesh being towards the top of the piston. Schrober et al. [16] conducted similar physical experiments combined with a CFD analysis. However, in their case, hydraulic oil was used as the liquid piston fluid to compress helium. The CFD results were validated successfully with physical experiments on the same system, whilst allowing for the analysis of the convective heat transfer values relating to helium undergoing compression.

The above studies mainly focused on the heat transfer between the fluids within the accumulator and the internal accumulator walls. The heat transfer between the external accumulator walls and the surrounding environment however also deserves attention. The latter was only partially experimented upon by Khaljani et al. [12] but was ignored by the remaining studies. In fact, in most numerical modelling approaches using CFD, the surrounding environment was replaced by a constant temperature wall boundary condition.

The present study deals with the CFD modelling of a liquid piston accumulator operating in a subsea environment during HPES system charging. The aim of this study is to apply a numerical modelling approach able to concurrently simulate the convective heat transfer processes on both the internal and external surfaces of an HPES accumulator when air is being compressed. The paper is organised as follows: initially the physical setup is explained, followed by a description of the CFD model and operational conditions being simulated. The numerical predictions from the

CFD model are then presented, together with derived values for the convective Heat Transfer Coefficient (HTC) for both the external and internal cylindrical surfaces of the accumulator. Two different approaches for deriving the internal HTC values are presented and discussed.

It should be noted that this study was part of the SAICOPES project currently being undertaken at University of Malta which aims to analyse the thermodynamic performance of pressure vessels undergoing compression cycles in a subsea environment.

2. Modelled HPES Accumulator

The HPES accumulator being modelled using CFD was assumed to consist of a vertically-oriented cylindrical pressure vessel with hemi-spherical end caps as shown in Figure 1. It was assumed that the accumulator is initially filled with air at an absolute pressure of 1 bar (i.e.: atmospheric pressure) which is gradually compressed up to around 100 Bar using a liquid piston. The latter is created through the injection of water at constant flowrate at the lower end of the accumulator. The geometric details of the accumulator are presented in Figure 1. Two different flowrates were simulated to investigate the influence of the hydraulic power delivered to the accumulator on the heat transfer phenomena. Table 1 shows the main geometrical and flow rate parameters used. It should be noted that the dimensions of the pressure vessel were based on the dimensions used by the SAICOPES project, albeit at a scale of 1:20 to save on computational requirements. The SAICOPES project pressure vessel dimensions were in turn based on standard offshore pipeline dimensions which were suggested for this particular use.

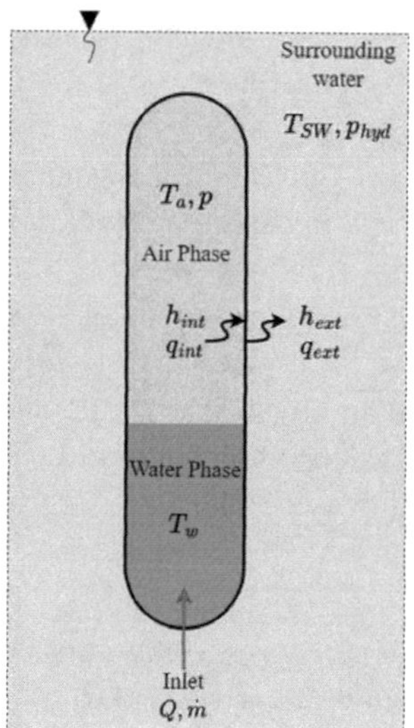

Figure 1. Schematic of CFD setup used to simulate an accumulator undergoing an HPES charge cycle under water. Not to scale

The accumulator was set at 50m below sea level, with an external hydrostatic pressure, p_{hyd}, of 590,942 Pa. This particular sea depth was chosen as this is the typical sea depth where current shallow water wind farms are being constructed. The initial air pressure within the pressure vessel was set to 1 bar (101,325 Pa), with the simulation time extended until the air in the accumulator reached 100 Bar.

Table 1. Accumulator Geometry and Modelled Flowrates

Parameter	Value
Accumulator diameter, D (m)	0.075
Cylindrical section length, L (m)	7.5
Accumulator wall thickness, t (mm)	2.525
Flowrate 1, Q_1 (l/min)	31
Flowrate 2, Q_2 (l/min)	15.5

3. Methods

3.1. CFD Mesh Setup

The CFD simulations were carried out using Ansys Fluent® 2023 R2. Figure 2 shows the geometry and measurement setup used for the CFD simulations, whilst Figure 3 shows different locations of the mesh setup used for the main simulations. The CFD simulations used a 2D axisymmetric model to help save on computational power. The axis of rotational symmetry was placed at the centre of the accumulator, with this represented with the dashed black line in Figure 2.

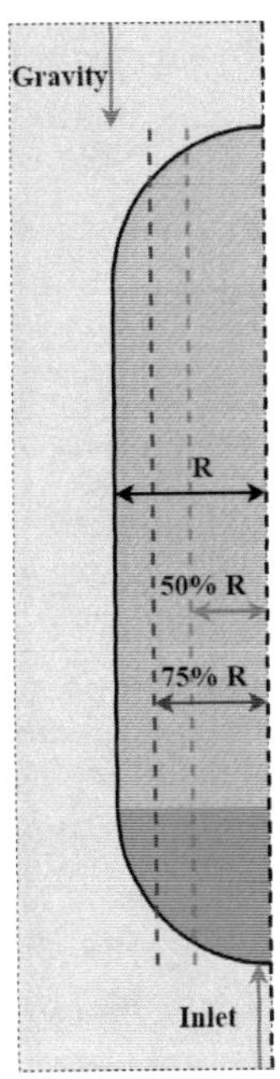

Figure 2. CFD geometry setup, including boundaries and measurement points. Not to scale

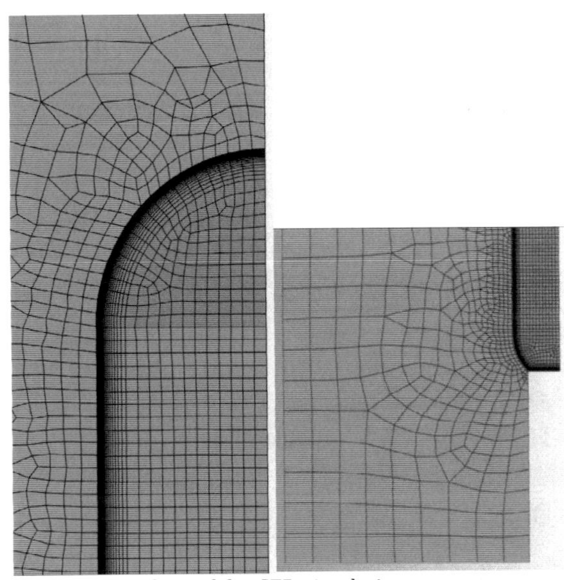

Figure 3. Mesh used for CFD simulations

The internal area representing the accumulator was split into three sections, representing the cylindrical section (shaded in green in Figure 2), as well as the upper and lower hemi-spherical ends (shaded orange and red respectively in Figure 2). A fourth section represented the external water surrounding the pressure vessel (shaded blue in Figure 2), with this set to be around 250 mm clearance around the accumulator. This value was set after initial simulations were carried out on similar model setups to ensure an adequate clearance around the pressure vessel. The accumulator walls were set directly via Ansys Fluent, and thus these were represented within the geometrical model as just a boundary line dividing the pressure vessel sections from the surrounding water section. A mesh size of 199,520 elements was chosen after a mesh sensitivity study was carried out. The criteria for the mesh sensitivity study were results within a 10% margin of error for pressure measurements and 1% error for temperature measurements.

The mesh included a wall inflation to help model the fluid boundary layer (both viscous and thermal) accurately. This was done by ensuring that the initial layer was small enough for the wall "y+" value to be lower than 1 throughout the length of both internal and external walls for the full duration of the compression cycle.

3.2. CFD Simulation Setup

The accumulator was set to a vertical orientation, with gravity enabled to ensure that natural convection was accurately represented within the model. The Volume of Fluid (VOF) multiphase model was used to accurately represent the interaction between the air and water phases within the accumulator. The turbulence model was set to use the k-ω SST turbulence model, as suggested by Ansys® documentation to aid with numerical stability [17].

Three materials were used: air, water and steel. Air was set to use the ideal gas law for the measurement of its density, whereas water used the Boussinesq formulation to calculate its density. All the remaining parameters were prescribed to the default parameters as set by Ansys Fluent®. The accumulator walls were modelled as a coupled wall with a finite thickness as described in Table 1. This enabled the use of inbuilt Ansys Fluent® functions to calculate the conductive heat transfer within the accumulator walls, without the requirement to have these modelled within the mesh.

The external water boundaries were fixed as pressure outlets, with backflow enabled to allow the simulation to represent the generated flow surrounding the accumulator due to natural convective heat transfer only, without any generated sea currents. The initial temperature, as well as the backflow temperature for the external pressure outlets and the surrounding sea water were set to 293.15 K. The accumulator inlet was modelled as a mass flow inlet at a temperature of 293.15 K and the two runs used with the flow rates input values as per Table 1. The SIMPLEC algorithm was selected since this recommended for simulations involving a gas undergoing compression [17].

3.3. Derived Data from CFD Postprocessing

Data for the relevant parameters was derived from the CFD model along four vertical lines within the accumulator, each at a different distance from the central axis: 0, 50, 75 and 100% R, as shown in Figure 2. It should be noted that 0 and 100% R denote the accumulator central axis and internal steel surface, respectively. The CFD model generated data for the parameters listed in Table 2 at each mesh node position along the particular vertical line at an interval of 3s. This provided an accurate representation of the distribution for each derived parameter as a function of time.

Table 2. Main parameters derived from CFD postprocessing

Parameters	Symbol	Location
Heat transfer flux	q_{int}, q_{ext}	100% R
Accumulator wall temperature	T_{wi}, T_{we}	100% R
Internal fluid temperature	T_0, T_{50}, T_{75}	$0, 50$ *and* 75% R
Phase	V_{wr}	0% R
Pressure	p	0% R

The external temperature of the surrounding water was also recorded as a mass-averaged value, "T_{sw}". Furthermore, mass-averaged data for the pressure and temperature of the air within the top hemisphere of the accumulator were obtained to give an overall indication of these two parameters within the accumulator at an interval of 1s. Note that the mass-average was only taken for the fluid within the upper hemisphere since during the simulation, this was the only location that was only filled with air. Had the cylindrical section also been included, the mass-averaged data would have included the water phase as well, skewing the results.

For the results section, computational results were taken at the following discrete time steps:

- V_W 10%: 6s for the 31 lpm flow rate, and 12s for the 15.5 lpm flow rate
- V_W 25%: 15s for the 31 lpm flow rate, and 30s for the 15.5 lpm flow rate
- V_W 50%: 30s for the 31 lpm flow rate, and 60s for the 15.5 lpm flow rate
- V_W 75%: 48s for the 31 lpm flow rate, and 96s for the 15.5 lpm flow rate
- V_W 90%: 57s for the 31 lpm flow rate, and 114s for the 15.5 lpm flow rate

4. Results

4.1. Pressure and Hydraulic Power

Figure 4 shows the resultant pressure within the accumulator as a function of time. It was then possible to calculate the accumulator inlet hydraulic power, which corresponds to the work being done on the accumulator air to compress it. The hydraulic work being done on the accumulator air was calculated assuming a hydraulic supply initially at atmospheric conditions using (1). The resultant power as a function of time is shown in Figure 5.

$$P = p \cdot Q \tag{1}$$

The resultant hydraulic power at the inlet, as expected, increased as the pressure increased since the inlet flow rate was kept constant throughout the compression cycle. This is distinct from the expected operation of this particular system since the inlet hydraulic power is to be kept constant throughout the compression cycle by reducing the flow rate at the inlet as the accumulator pressure increases. Since the CFD simulations used constant flow rates set at the highest values that would be expected at the start of compression cycles, the CFD simulations represent a worst-case scenario where the system is charged faster than intended, thus leading to the possibility that the accumulator is not allowed to cool down fast enough to enable isothermal operation.

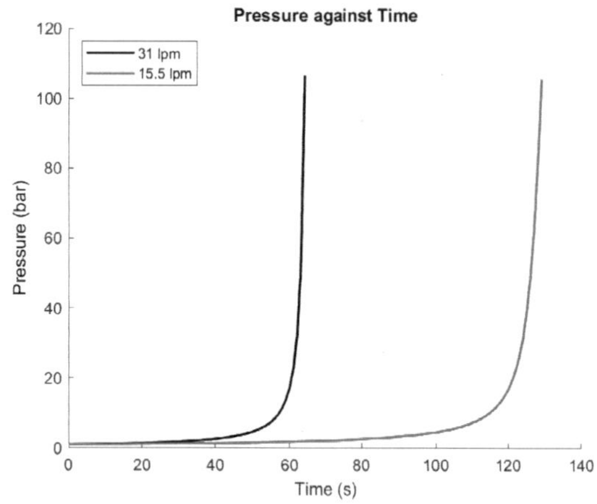

Figure 4. Accumulator pressure across compression cycle

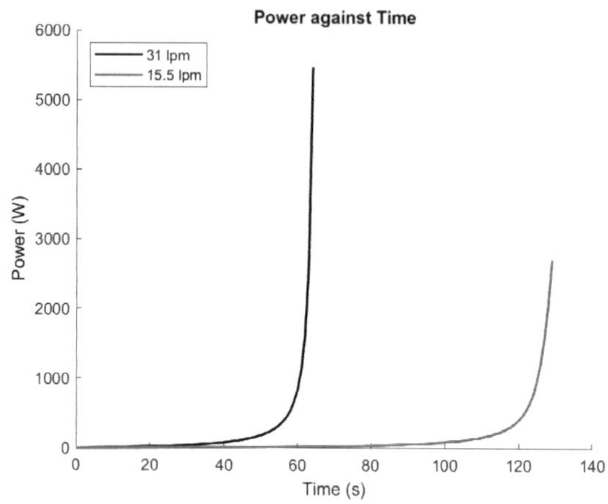

Figure 5. Accumulator inlet hydraulic power

4.2. Internal Temperatures

Figure 6 shows the air and water temperatures within the accumulator at different time stamps. The temperature variation is shown as a function of the vertical position within the accumulator for the three different measurement locations as shown in Figure 2. Note that the surrounding water temperature, T_{SW}, is added for reference in Figure 6.

4.3. Internal HTC

Figure 7 shows the convective HTC for the air phase only as calculated using (2). In this case, the HTC is calculated for every mesh node on the accumulator wall by using the heat flux and the internal wall surface temperature at that particular mesh node. As for the fluid temperature, this was set to the air temperature at the accumulator centre line (0%R, T_0) at the same vertical position as the mesh node on the accumulator wall.

$$h_{int0} = \frac{q_{int}}{T_{wi} - T_0} \qquad (2)$$

Figure 8 shows the convective HTC for the air phase only as calculated using (3). In this case, the fluid temperature used to calculate the convective HTC uses the average temperatures of the fluid as measured at the three different internal measurement points (0, 50 and 75%R) at the node's particular vertical position. This thus uses a "bulk"-averaged temperature, T_{BULK}, of the fluid within the accumulator as per (4.

$$h_{intB} = \frac{q_{int}}{T_{wi} - T_{BULK}} \qquad (3)$$

$$T_{BULK} = \frac{T_0 + T_{50} + T_{75}}{3} \qquad (4)$$

A comparison between the convective HTC as calculated using Equations (2) and (3) is shown in Figure 9. The discrepancy between the two HTCs is due to the temperature distribution within the accumulator, with the data taken on the vertical lines closer to the wall falling within the thermal boundary layer. The considerable discrepancy shown in Figure 9 between the two above approaches shows that special care should be taken when deriving the transient HTC values from CFD.

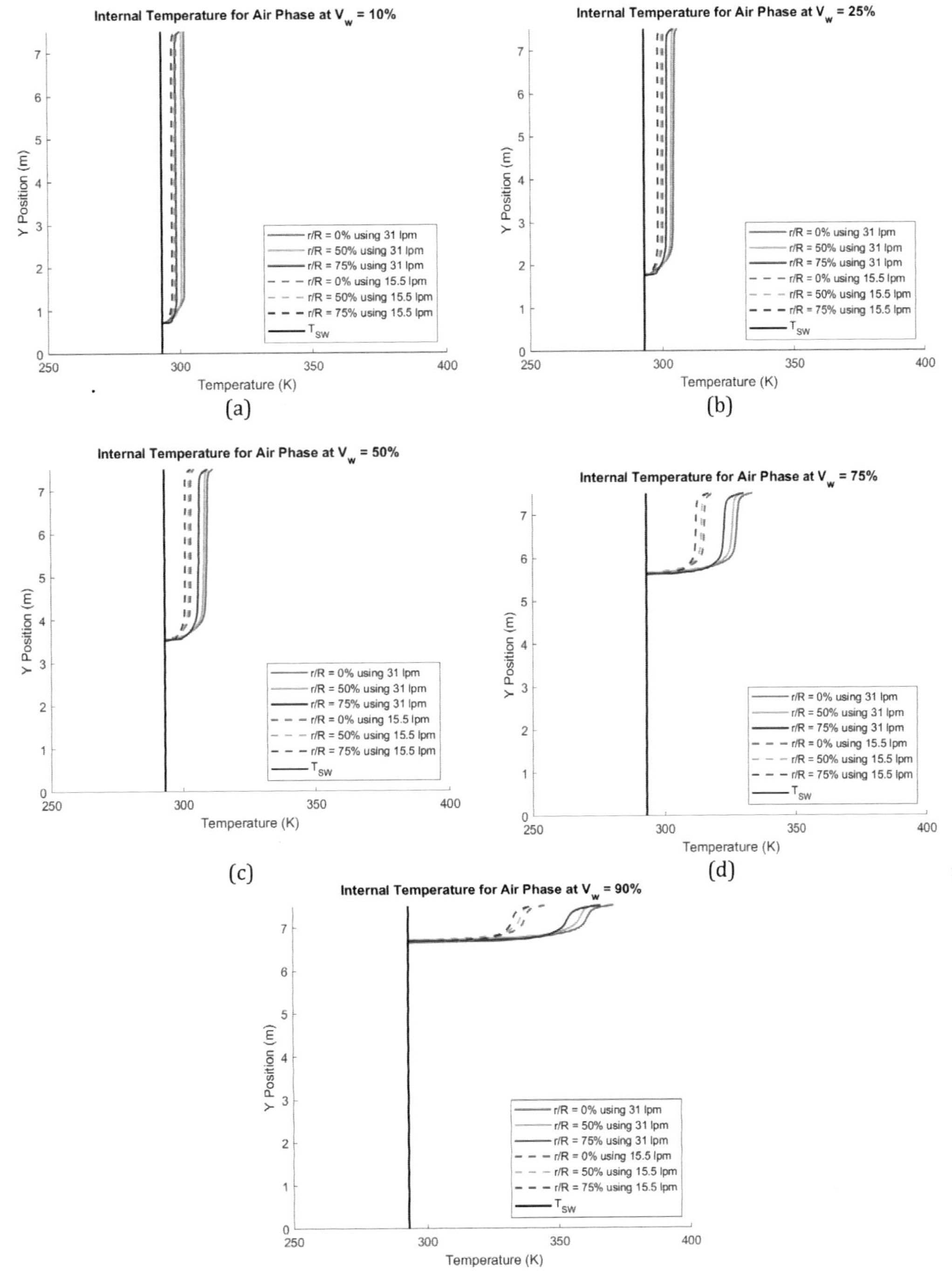

Figure 6. Air temperature within accumulator at: a) 10%, b) 25%, c) 50%, d) 75% and e) 90% water volume fraction

The resultant convective HTC gives a good indication of the heat transfer present within the accumulator. Of particular interest is the fact that the HTC increases as the air pressure and temperature within the accumulator increase, which is in line with that found by Wieberdink et al. and Schober et al. [15, 16]. However, during both compression cycles, the HTC remains relatively and similar to what would be expected in natural convection.

Another particular aspect was the fact that the internal convective HTC remains relatively constant along the vertical position within the accumulator for a large part of the compression cycle except for regions close to the air-water boundary and the top hemisphere. However, towards the end of the cycle where the accumulator air is compressed to a much smaller volume, the internal convective HTC does vary noticeably with position. This is possibly due to induced flow turbulence. However, further investigations using more comprehensive CFD methods, such as Large-Eddy Simulations (LES) are necessary to better understand the reason for such variations in HTC values along the vertical position.

4.4. External HTC

Figure 10 shows the resultant convective HTC pertaining to the resultant natural convection at the external wall of the accumulator to the surrounding water. In this case, the external convective HTC was calculated using (5), with the fluid temperature, T_{SW}, set to 293.15 K. As noted from the results, the external convective HTC is much higher than the internal convective HTC, with values falling within the ranges expected for natural convection for water. This further highlights the importance of having the accumulator surrounded by water to help maintain isothermal conditions, which then leads to optimal operation of the HPES system.

$$h_{ext} = \frac{q_{ext}}{T_{we} - T_{SW}} \tag{5}$$

4.5. Polytropic index

Figure 11 shows the resultant polytropic index for the two compression process presented. In both cases, the polytropic index was calculated at 1s intervals using (6).

$$n = \frac{\ln(p_1) - \ln(p_2)}{\ln(v_2) - \ln(v_1)} \tag{6}$$

As shown for both cases, the polytropic exponent is close to unity for most of the charging cycle, demonstrating that the thermodynamic compression process of the air inside the accumulator is quasi-isothermal. Also of note is the fact that the initial phase of the compression cycle leads to instances where the polytropic index is less than unity. This is possibly due to the initial variation in pressure and volume between iterations, coupled with the high heat transfer between the air and the surrounding subsea environment which is a result of the large surface area between the two as the liquid piston is still in the lower levels of the accumulator.

This indicates that if the inlet flow rate is to be modulated to maintain constant power, the system would be expected to operate even closer to isothermal conditions.

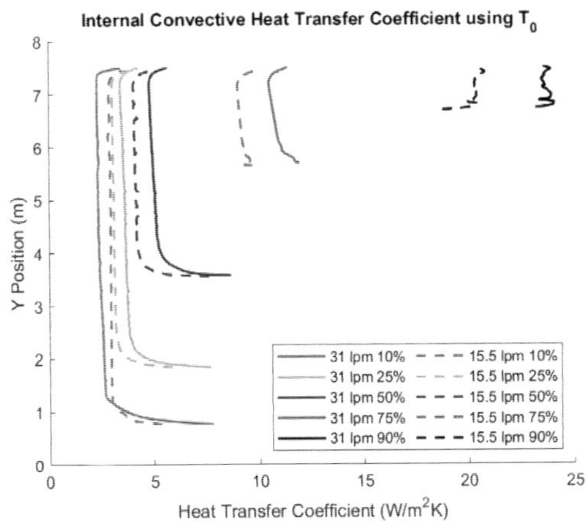

Figure 7. Internal convective HTC calculated using T_0

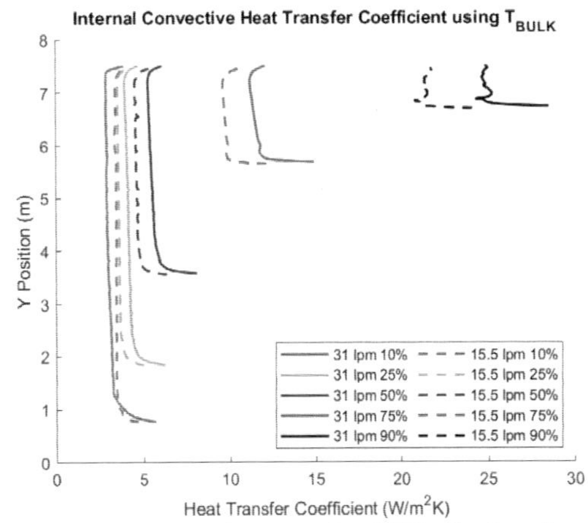

Figure 8. Internal convective HTC calculated using T_{BULK}

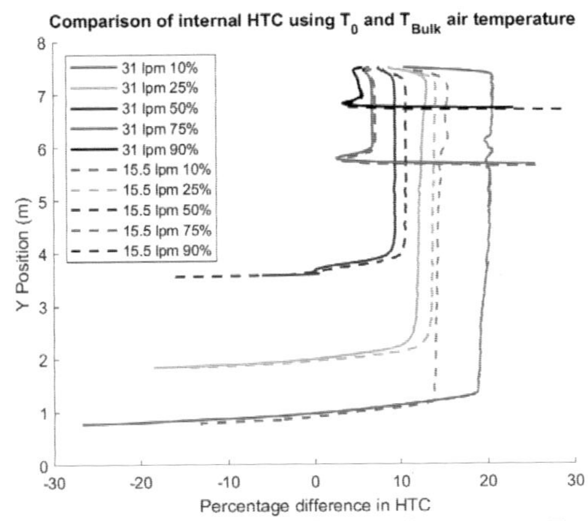

Figure 9. Comparison of internal HTC using T_0 and T_{BULK}

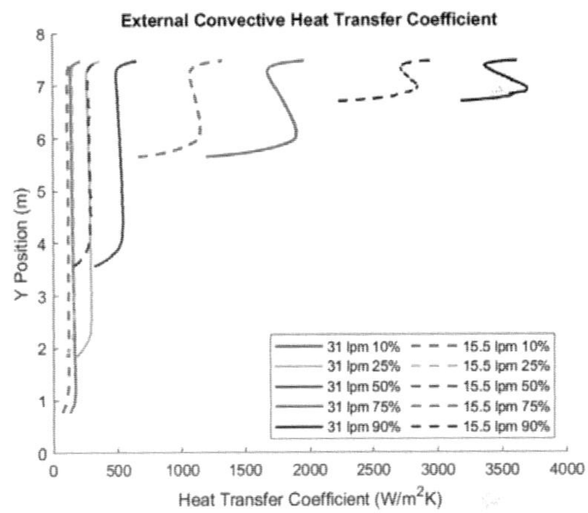

Figure 10. External convective HTC

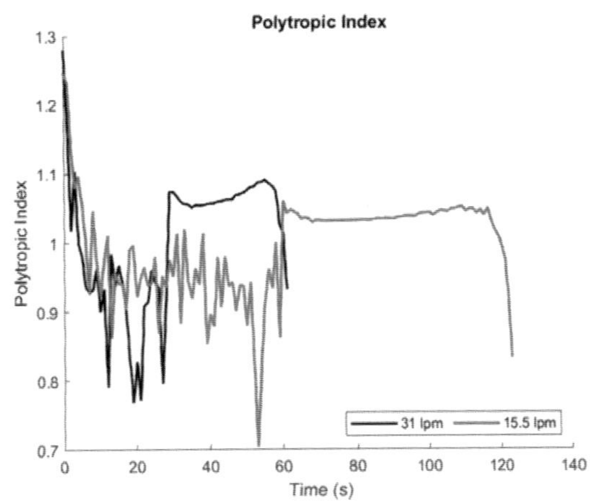

Figure 11. Compression processes polytropic index

5. Conclusion

From the CFD simulations, it was noted that the internal convective HTC increases as the system pressure and temperature increase, going from about 2.5 W/m²K to 25 W/m²K across the entire charging cycle of the HPES accumulator. This was also noted for the external convective HTC, with values increasing from around 150 W/m²K to 3750 W/m²K. Such HTC values may thus be used to help understand better the system as it is undergoing compression, leading to numerical models which can represent the system accurately.

It should be noted that the higher the pressure and temperature within the accumulator, the higher the fluctuations of the obtained HTCs using CFD. This is possibly due to internal flow turbulence, particularly towards the end of the compression cycle as the air is compressed to a small fraction of volume at the top of the accumulator. Further studies involving the combined use of lab experiments and CFD are thus recommended to examine this phenomenon and how this effects the heat transfer processes. Future work should also consider repeating the analysis for a wider range of pressure ratios and initial pressures. Finally, it will also be essential to investigate how changing the water flow to the accumulator from one that is time invariant to a variable flowrate that maintains a constant hydraulic power input will impact the HTC values. These would however require much more computational time when compared to simulations in which a constant flowrate is prescribed at the inlet.

Acknowledgments

The presented research forms part of project SAICOPES (Subsea Air Isothermal Compression using Offshore Pipelines in Long Duration Energy Storage Applications) funded by Xjenza Malta under the Research Excellence Programme. Project Reference: REP-2023-004.

References

[1] Badal, F. R., Das, P., Sarker, S. K., and Das, S. K., 'A survey on control issues in renewable energy integration and microgrid', *Protection and Control of Modern Power Systems*, vol. 4, no. 1, p. 8, Dec. 2019, doi: 10.1186/s41601-019-0122-8.

[2] Kumar, G. V. B., Sarojini, R. K., Palanisamy, K., Padmanaban, S., and Holm-Nielsen, J. B., 'Large Scale Renewable Energy Integration: Issues and Solutions', *Energies (Basel)*, vol. 12, no. 10, p. 1996, May 2019, doi: 10.3390/en12101996.

[3] Sinsel, S. R., Riemke, R. L., and Hoffmann, V. H., 'Challenges and solution technologies for the integration of variable renewable energy sources—a review', *Renew Energy*, vol. 145, pp. 2271–2285, Jan. 2020, doi: 10.1016/j.renene.2019.06.147.

[4] Ould Amrouche, S., Rekioua, D., Rekioua, T., and Bacha, S., 'Overview of energy storage in renewable energy systems', *Int J Hydrogen Energy*, vol. 41, no. 45, pp. 20914–20927, Dec. 2016, doi: 10.1016/j.ijhydene.2016.06.243.

[5] Behabtu, H. A. *et al.*, 'A Review of Energy Storage Technologies' Application Potentials in Renewable Energy Sources Grid Integration', *Sustainability*, vol. 12, no. 24, p. 10511, Dec. 2020, doi: 10.3390/su122410511.

[6] Yekini Suberu, M., Wazir Mustafa, M., and Bashir, N., 'Energy storage systems for renewable energy power sector integration and mitigation of intermittency', *Renewable and Sustainable Energy Reviews*, vol. 35, pp. 499–514, Jul. 2014, doi: 10.1016/j.rser.2014.04.009.

[7] Odukomaiya, A., Abu-Heiba, A., Graham, S., and Momen, A. M., 'Experimental and analytical evaluation of a hydro-pneumatic compressed-air Ground-Level Integrated Diverse Energy Storage (GLIDES) system', *Appl Energy*, vol. 221, pp. 75–85, Jul. 2018, doi: 10.1016/j.apenergy.2018.03.110.

[8] Camargos, T. P. L., Pottie, D. L. F., Ferreira, R. A. M., Maia, T. A. C., and Porto, M. P., 'Experimental study of a PH-CAES system: Proof of concept', *Energy*, vol. 165, pp. 630–638, Dec. 2018, doi: 10.1016/j.energy.2018.09.109.

[9] Chen, H., Wang, H., Li, R., Sun, H., Ge, G., and Ling, L., 'Experimental and analytical investigation of near-isothermal pumped hydro-compressed air energy storage system', *Energy*, vol. 249, p. 123607, Jun. 2022, doi: 10.1016/j.energy.2022.123607.

[10] Buhagiar, D., Sant, T., Farrugia, R. N., Aquilina, L., Farrugia, D., and Strati, F. M., 'Small-scale Experimental Testing of a Novel Marine Floating Platform with Integrated Hydro-pneumatic Energy Storage', *J Energy Storage*, vol. 24, p. 100774, Aug. 2019, doi: 10.1016/j.est.2019.100774.

[11] Gouda, E. M., Neu, T., Benaouicha, M., Fan, Y., Subrenat, A., and Luo, L., 'Experimental and numerical investigation on the flow and heat transfer behaviors during a compression–cooling–expansion cycle using a liquid piston for compressed air energy storage', *Energy*, vol. 277, p. 127622, Aug. 2023, doi: 10.1016/j.energy.2023.127622.

[12] Khaljani, M., Vennard, A., Harrison, J., Surplus, D., Murphy, A., and Mahmoudi, Y., 'Experimental and modelling analysis of efficiency enhancement in a liquid piston gas compressor using metal plate inserts for compressed air energy storage application', *J Energy Storage*, vol. 43, p. 103240, Nov. 2021, doi: 10.1016/j.est.2021.103240.

[13] Mutlu, M. and Kiliç, M., 'Effects of piston speed, compression ratio and cylinder geometry on system performance of a liquid piston', *Thermal Science*, vol. 20, no. 6, pp. 1953–1961, 2016, doi: 10.2298/TSCI140926146M.

[14] Zhang, C., Li, P. Y., Van de Ven, J. D., and Simon, T. W., 'Design analysis of a liquid-piston compression chamber with application to compressed air energy storage', *Appl Therm Eng*, vol. 101, pp. 704–709, May 2016, doi: 10.1016/j.applthermaleng.2016.01.082.

[15] Wieberdink, J., Li, P. Y., Simon, T. W., and Van de Ven, J. D., 'Effects of porous media insert on the efficiency and power density of a high pressure (210 bar) liquid piston air compressor/expander – An experimental study', *Appl Energy*, vol. 212, pp. 1025–1037, Feb. 2018, doi: 10.1016/j.apenergy.2017.12.093.

[16] Schober, M., Deichsel, M., and Schlucker, E., 'CFD-Simulation and Experimental Validation of Heat Transfer in Liquid Piston Compressors', in *12th International Conference on Heat Transfer, Fluid Mechanics and Thermodynamics*, Costa del Sol, 2016.

[17] Ansys Inc., *Ansys Fluent User's Guide 2023R2*. 2023.

Thermodynamic and Economic Analysis of a Trigenerative Adiabatic Compressed Air Energy Storage

Sunku Prasad Jenne[1], Chenghao Li[1], Xuecen Zhang[1], Helena Navarro[2], Jihong Wang[1], Wei He[3], Joe Spencer[4], Yulong Ding[2]

[1]School of Engineering, University of Warwick, UK
[2]Birmingham Centre for Energy Storage, School of Chemical Engineering, University of Birmingham, Birmingham B15 2TT, UK
[3]Department of Engineering, King's College London, UK
[4]Department of Electrical Engineering and Electronics, University of Liverpool, Liverpool L693GJ, UK
*E-mail: sunku-prasad.jenne@warwick.ac.uk

Abstract. Energy storage is indispensable for decarbonising power systems with high penetration of fluctuating renewable generation (e.g., wind, solar). In this study, an innovative trigeneration adiabatic compressed air energy storage (Tri-ACAES) system that produces heating, cooling, and electricity is proposed. The Tri-ACAES system can absorb more wind power than an ACAES system. The Tri-ACAES can work in heating and cooling mode when renewable generation is too low to charge the ACAES system due to the off-design characteristics of the compressor. The results reveal that the round-trip efficiency and turbine power output of the proposed Tri-ACAES system are enhanced by up to 5.8% and 33%, respectively than conventional ACAES system. Further, the proposed system reduced the cost of thermal energy storage by up to 60% than conventional ACAES system.

1. Introduction

Compressed Air Energy Storage (CAES) is one of the potential options for large scale, medium duration energy storage with unique features like scalability, cost-effectiveness (2-50 $/kWh), and environmental sustainability (1). CAES operates on the principle of converting surplus electricity into compressed air during periods of low electricity demand. The compressed air is then stored in underground reservoirs or aboveground artificial containers. When electricity demand peaks, the stored compressed air is expanded in turbines to generate electricity, thus contributing to grid stabilization, and meeting peak load requirements. Currently, two Diabatic-CAES systems are in operation by burning natural gas, namely Huntorf (321 MW) and McIntosh (110 MW) CAES plants, with round trip efficiencies (RTE) of 42% and 54%, respectively. The Huntorf plant consumes 0.8 kWh of electricity (for compression) and 1.6 kWh of natural gas (for combustion) to generate 1 kWh of electricity. In contrast, the McIntosh plant integrates a recuperator to recover some residual heat, using 0.69 kWh of electricity and 1.17 kWh of gas to produce 1 kWh of electricity (2). On the other hand, Adiabatic-CAES (ACAES) systems capture the compression heat for storage and utilize it during expansion without the need to burn fossil fuels.

The deployment of ACAES is accelerating, after a decade-long development globally (3). However, the reported RTE of the ACAES system lies in the range of 60-75%, which is not competitive with pumped hydro (65-85%) and electrochemical battery (85-95%) (1)(4). Further, the CAES system has less operation flexibility than electrochemical battery and pumped hydro due to the poor performance of the compressor under off-design conditions. Therefore, a new ACAES configuration is proposed to increase RTE and the operation flexibility of the ACAES system in the present study. A thermodynamic, and economic analysis is performed on ACAES and modified (proposed) ACAES systems. Further, the performance of newly proposed ACAES configuration is compared with the traditional ACAES system in terms of power capacity, RTE and power capital cost.

2. System description

In a typical 2-stage ACAES system (shown in Fig. 1a, referred as configuration 1/ ACAES), the compressor outlet temperatures fall within the range of 275-400°C for the maximum cavern pressures ranging from 50-170 bar. Recent studies reported that the maximum acceptable outlet temperature of compressor is about 600°C (5)(6)(7)(8). This is notably higher than the typical limit of 177°C for conventional air compressors. However, such elevated outlet temperatures are more commonly associated with aero-engine/power-generating gas turbines (9). A modified ACAES system as shown in Fig. 1b is proposed based on the safe operating outlet temperature of the compressor, referred as configuration 2/ Tri-ACAES. The present study considers 600°C as the maximum safe operating outlet temperature of the compressor (7). During charging (1-2-3-4-5-6-7), air is compressed in the first compressor and then passed through second compressor without intercooling. The temperature of the air at the outlet of second compressor is raised to maximum operating limit. The compressor 1 and 2 have equal pressure ratios ($r_{c1} = r_{c2}$) and the outlet pressure of compressor 2 is fixed to satisfy the maximum operating temperature limit of compressor. The hot air is cooled to ambient temperature in the Packed Bed Thermal Energy Storage (PBTES) before sending it to the third compressor. The third compressor raises the air pressure to the maximum pressure of the air storage reservoir. The heat generated in the third compressor is recovered, stored and can be utilised for space/ district heating applications. The cooled air is then stored inside air reservoir. During discharging (8-9-10-11-12), the cavern air is passed through the PBTES at discharge pressure and is expanded in the turbines after extracting the stored thermal energy from the PBTES.

In configuration 2, compressor 3 is responsible for delivering the thermal energy for district heating, which can be operated independently if needed, regardless of the charging and discharging. Compressors are most efficient at their design point (10). For the CAES, operating the compressors at design point is difficult due to fluctuating wind power. Deviation from design point operation leads to reduced efficiency. Therefore, compressors are less flexible for operating at maximum efficiency (only 5% margin of design point under constant speed). Therefore, an air refrigeration cycle (ARC) (5-6'-7'-8') is constituted using compressor 3 to capture very low amount (10 – 30% of rated power of whole compressors) of excess renewable electricity and store it in the form of heat and cold, as shown in Fig. 1b. The ARC is widely used in aircraft and cryogenic fields (11). The ARC comes into operation when the excess renewable power is not enough to run the all the compressors near to their peak effiecncy point. This strategy helps to increase the operational range of the compressor train of the CAES system. The following are the advantages of configuration 2 over configuration 1.

- The inlet temperature of air to the turbine is higher compared to configuration 1, which leads to high work output from the turbines.

- Two PBTESs in configuration 1 are replaced with a single PBTES, which helps to reduce the TES cost.
- Unlike the configuration 1, the PBTES of configuration 2 is only designed for discharge pressure (i.e., for lower pressure but higher temperature), which saves storage tank mass.
- In configuration 2, part of the charging power is recovered in the temperature range of 90-180°C for the maximum cavern pressures of 50-100 bar and can be stored and transported to district heating networks.
- Configuration 2 can capture a lower percentage of rated compressor power and disseminate it into heat and cold energy.
- Configuration 2 can absorb more wind power and thereby reducing the wind curtailment cost than configuration 1.

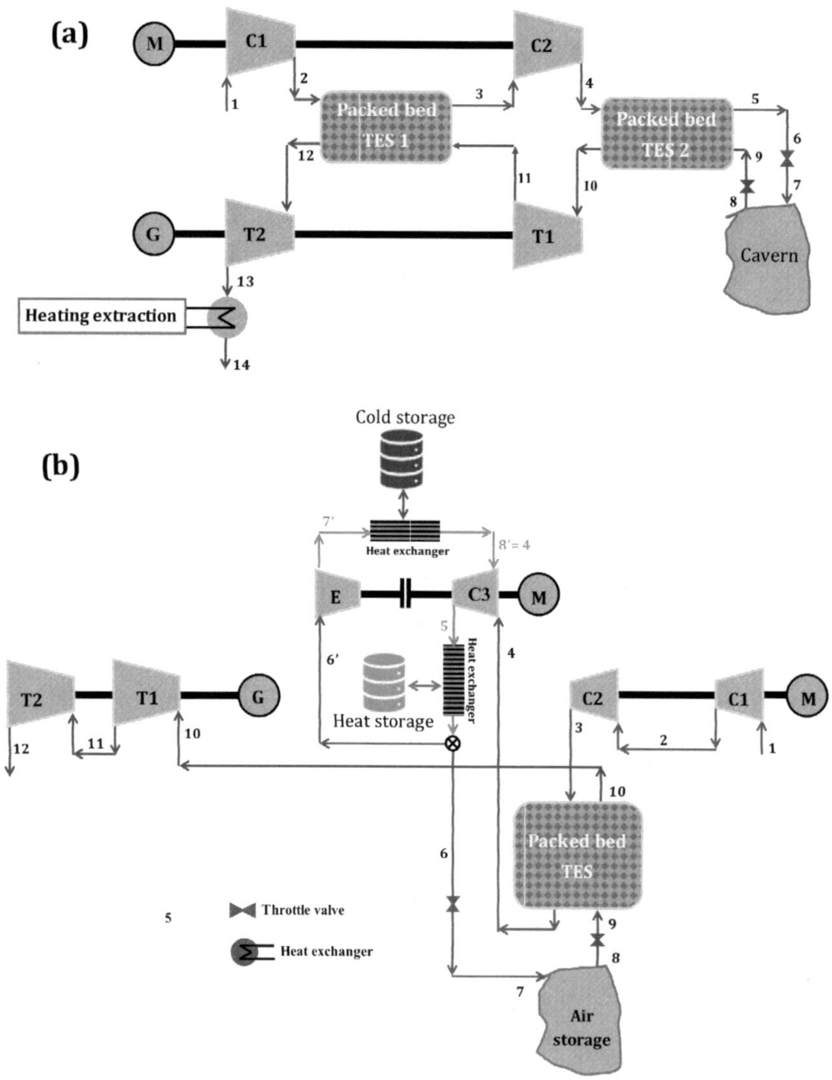

Figure. 1 (a) Typical 2-stage ACAES system (configuration 1), (b) modified ACAES system (configuration 2/ Tri-ACAES)

3. Modelling of CAES systems

3.1 Compressor and turbine

During charging, air is always compressed to a maximum cavern pressure and throttled to the current cavern pressure, resulting in exergy losses and a reduction in efficiency. During discharging, air is throttled to the lower limit of the cavern pressure before expanding in the turbine. This methodology is best suited if the compressor has limited operational flexibility. Sciacovelli et al. (7) showed that the compressors consumed 7.6% more power when the low and high pressure compressor's compression ratios are 10% and 22% lower than the design value. This is due to minimum isentropic efficiency of the compressors under off-design conditions at the beginning of the charging process. Therefore, the compressor is sized for the maximum cavern pressure and then air is throttled to achieve a constant compression ratio as the cavern fills. The steady state equations for compressor and turbine are shown in Table 1. The polytropic efficiency of compressor and turbine are estimated based on the pressure ratio using the correlations presented by Mersch et al. (2023) (13).

Table 1 Steady state model equations for compressor and turbine (13)

Compressor	Outlet temperature of compressor $T_{c,out} = T_{c,in}\left(\frac{p_{c,out}}{p_{c,in}}\right)^{\frac{\gamma-1}{\gamma\times\eta_{p,comp}}}$	(1)
	Polytropic efficiency of compressor $\eta_{p,comp} = 0.91 - \frac{\left(\frac{p_{c,out}}{p_{c,in}}\right)-1}{300}$	(2)
	Power consumed by compressor $W_{C_{1,2,3,\ldots,n}} = \dot{m}_{ch} C_p \left(T_{out,C_{1,2,3,\ldots,n}} - T_{in,C_{1,2,3,\ldots,n}}\right)$	(3)
Turbine	Outlet temperature of turbine $T_{t,out} = T_{t,in}\left(\frac{p_{t,out}}{p_{t,in}}\right)^{\frac{(\gamma-1)\times\eta_{p,turb}}{\gamma}}$	(4)
	Polytropic efficiency of turbine $\eta_{p,turb} = 0.9 - \frac{\left(\frac{p_{t,in}}{p_{t,out}}\right)-1}{250}$	(5)
	Power generated by turbine $W_{T_{1,2,3,\ldots,n}} = \dot{m}_{disch} C_p \left(T_{in,T_{1,2,3,\ldots,n}} - T_{out,T_{1,2,3,\ldots,n}}\right)$	(6)

The cost correlations for the compressor and turbine are shown in Table 2. The cost correlations for turbomachinery based on the power rating are used and applicable for compressor outlet temperatures up to 400°C with steel as construction material. In both the configurations, the compressor outlet temperatures fall within this limit, except for the second compressor in Configuration 2, which reaches an outlet temperature of 600°C. Due to this higher temperature, specialized materials are required for this compressor. However, no cost correlations/ cost data are available for such compressors. Therefore, we used the same cost correlation for the second compressor in Configuration 2 due to the lack of available cost correlations. It is important to note that the actual cost of this compressor may be higher due to the need for specialized materials. A Chemical Engineering Plant Costing Index (CEPCI) is used to account for inflation and all the costs are calculated for the year 2023. An electric drive cost of 120 $/kW is considered in the analysis for both compressor and turbine (14).

Table 2 Cost correlations for compressor and turbine (15)(16)

Component	Correlation	Reference year (CEPCI = 521.9)	Year considered (CEPCI = 800.8)
Compressor	$C_{comp}(\$) = 7900(Power\, in\, kW)^{0.62}$	2009	2023
Turbine	$C_{turb}(\$) = 1100(Power\, in\, kW)^{0.81}$	2009	2023

3.2 Sizing of the PBTES

A cylindrical packed bed is considered for the analysis. The pressure vessel is filled with spherical TES pellets. The pressure vessel is made of cast steel and stainless steel for configurations 1 and 2, respectively.

Mass of TES material required is given by Eq. 7.

$$M_{TES} = (t_{ch} \times \dot{m}_{air}) \left(\frac{C_{P\,air_{avg}}}{C_{P\,TES_{avg}}} \right) \times \beta_{factor} \tag{7}$$

The mass overrating factor (β_{factor}) is used to improve the storage efficiency, which is set to 1.6 in the present study. The diameter of the packed bed is calculated by Eq. 8.

$$D_b = \left(\frac{4 \times V_{total}}{\pi \times AR} \right)^{\frac{1}{3}} \tag{8}$$

The total inner volume of the packed bed is given by Eq. 9

$$V_{total} = \frac{V_{TES}}{(1-\varepsilon)} \tag{9}$$

The volume of the TES material is given by Eq. 10

$$V_{TES} = \frac{M_{TES}}{\rho_{TES}} \tag{10}$$

The aspect ratio (AR), defined as the height-to-diameter ratio of the packed bed, is given in Eq. 11.

$$AR = \frac{H}{D_b} \tag{11}$$

Minimum wall thickness of the packed bed pressure vessel for bearing internal pressure given by Eq. 12.

$$t_{min} = \frac{P_{max} OD}{2(S + P_{max})} \tag{12}$$

The allowable stress (S) of the pressure vessel material is considered based on the maximum operating temperature. The S value of carbon steel is considered as 117 MPa for the configuration 1. In the configuration 2, stainless steel is used for the pressure vessel and the S value of 106, 103, and 81 MPa are used in the design at 500, 550, and 600°C, respectively.

The packed bed thermal energy storage (PBTES) cost includes the cost of TES material (magnesia) and the cost of the pressure vessel (designed to withstand turbine inlet pressure). A recent study (13) found that the pressure vessel and storage material contributed only 15% of the total system cost, while turbomachinery accounted for a significant 84%. Therefore, the costs of thermal insulation, foundation, piping, construction, and maintenance were not included in the present analysis. Since the primary objective is to demonstrate the advantages of the proposed Tri-ACES system over conventional ACAES, excluding the less significant costs is justified. The thermophysical properties and the cost of materials used for the PBTES are presented in Table 3.

Table 3 Thermo-physical properties and cost of PBTES materials and steel (13)(17)

Material	Density (kg/m^3)	Specific heat (J/kg-K)	Thermal conductivity (W/m-K)	Cost ($/kg)
Magnesia	3581	960	37	0.5
Stainless steel (processed)	7870	500	16.3	4
Carbon steel	7840	490	43	0.8

3.3 Heat exchanger

The cost of the heat exchanger (or after cooler) of compressor 3 of the configuration 2 is given by Eq. 13 (15). Water inlet temperature and mass flow rate of 15°C and 36 kg/s are considered for calculating the heat exchange area using Eq. 14. Also, an overall heat transfer coefficient and effectiveness of 500 W/m^2-K and 0.8 are assumed. A detailed analysis for heat exchanger cost estimation is not performed due to its negligible contribution to the total cost of the system (~0.5 $/kW power delivered).

$$C_{HEX} = 16000\left(\frac{A_{HEX}}{100}\right)^{0.6} \tag{13}$$

$$Q_{HEX} = U_{overall} A_{HEX} LMTD \tag{14}$$

3.4 Performance parameters

A pressure recovery factor, δ_p (Eq. 15) is defined and the performance parameters are evaluated for various values of δ_p. The present study investigates the effect of operating pressure range of the air reservoir and the safe operating outlet temperature of the compressor on the performance of configurations 2. Also, the performance of configuration 2 is compared with the configuration 2. The operating pressure range of the air reservoir is varied from 23-50 bar (δ_p = 0.46) to 143-170 bar (δ_p = 0.8412) while keeping the transferred air mass constant. In this context, the pressure range of 23-50 bar implies that the compressor's outlet pressure is set at 50 bar, while the turbine's inlet pressure is 23 bar. For the parametric study, three safe operating outlet temperatures for compressor 2 in configuration 2 are chosen: 500°C, 550°C, and 600°C.

$$\delta_p = \frac{inlet\ pressure\ to\ turbine}{outlet\ pressure\ of\ compressor} \tag{15}$$

Shaft to shaft RTE is defined as the ratio of energy input during charging to the energy output during discharging, is given by Eq. 16.

$$RTE_{ACAES} = \frac{\left(W_{T_1} + W_{T_2} + W_{T_3} + ... + W_{T_n}\right)_{actual} \times t_{disch}}{\left(W_{C_1} + W_{C_2} + W_{C_3} + ... + W_{C_n}\right)_{actual} \times t_{ch}} \tag{16}$$

Power capital cost, defined as a total capital cost to the discharge power is used as an economic indicator (13), given by Eq. 17.

$$C_{PC} = \frac{C_{total} = C_{compressor} + C_{turbine} + C_{PBTES} + C_{heat\,exchanger}}{\dot{W}_{el,turb,1} + \dot{W}_{el,turb,2}} \tag{17}$$

For modelling, the cavern size, charging, and discharge conditions are adopted from Huntorf CAES plant, as shown in Table 4. Further, the heat loss from the PBTES is not considered and the turbine inlet temperature is assumed to be equal to the compressor outlet temperature.

Table 4 Huntorf operating conditions (18)

Variable	Value
Cavern volume, m3	310,000
Charging flow rate, kg/s	108
Discharging flow rate, kg/s	417

4. Results and discussion

4.1 RTE, power capacity, and power capital cost

In this section, the results of configurations 1 and 2 are compared under steady-state conditions. The RTE, turbine power, capital cost of turbomachinery, PBTES and power capital cost are compared.

The RTE of configurations 1 and 2 is plotted in Figure 2 with respect to the pressure recovery factor (δ_p). For configuration 1, the RTE is increased with pressure recovery factor due to reduced throttling losses and the increased compressor outlet temperatures. On the other hand, the RTE of configuration 2 is increased and reaches a maximum limit and decreases thereafter with respect to δ_p for the given outlet temperature limit of compressor 2. The RTE of configuration 2 reached a maximum value of 76.7%, 78.8%, 80.4% at δ_p of 0.663, 0.663, and 0.7 for the compressor outlet temperature limits of 500, 550, and 600°C, respectively. The RTE of configuration 2 decreased after certain value of δ_p due to the increase in the discarded heat of compressor 3 (refer to Figure 6). Further, configuration 2 showed superior RTE than configuration 1 (both 2- and 3-stage) up to specific value of δ_p only. It is observed that configuration 2 is affordable up to a certain level of discarded heat of compressor 3, beyond which configuration 2 yields lower RTE than configuration 1. The shaded area in Figure 2 is not recommendable for configuration 2 due to its lower RTE. In the unshaded area, configuration 2 showed a maximum RTE gain of 3.91%, 5.08%, and 5.82% than 2-stage configuration 1 at δ_p = 0.46. With increase in δ_p, the gain in RTE decreases and becomes negative in shaded area.

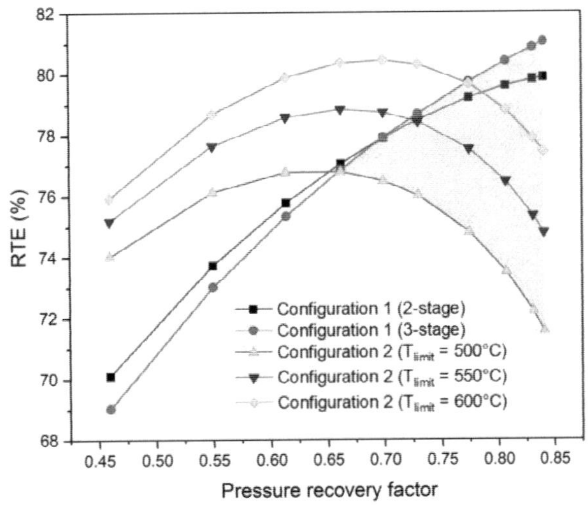

Figure 2. Comparison of RTE for configurations 1 and 2

Figure 3 shows the turbine discharge power and the gain in turbine discharge power for configurations 1 and 2. It is observed that configuration 2 boosted the turbine discharge power by a maximum of 18, 26, and 33% at δ_p = 0.46 for the compressor outlet temperature limits of 500, 550, and 600°C, respectively. Like RTE, the increase in turbine discharge power decreases with δ_p.

Overall, the proposed Tri-ACAES configuration boosts the RTE and turbine discharge power significantly at lower values of δ_p. Such a configuration is especially useful for achieving higher power capacity for the given air reservoir volume.

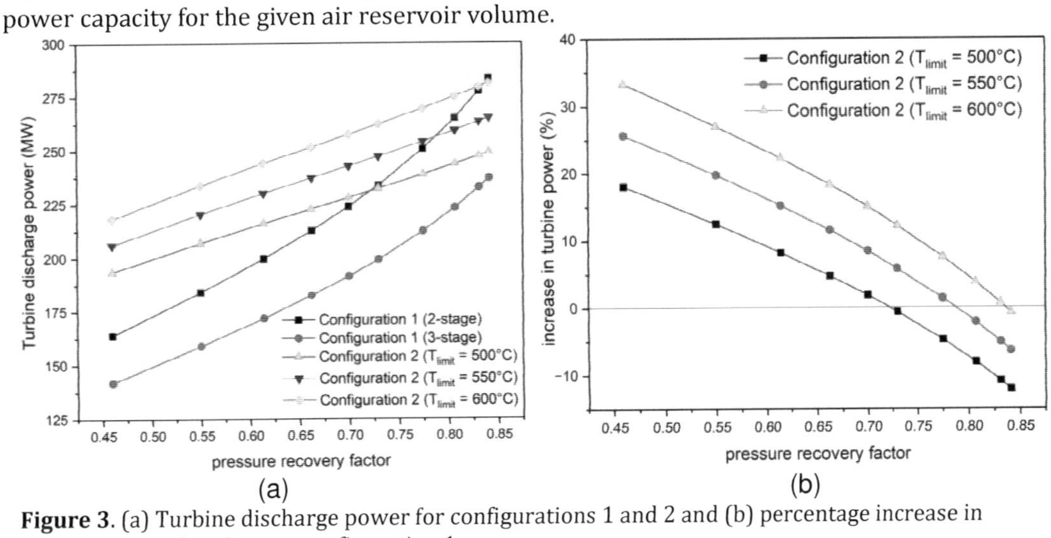

Figure 3. (a) Turbine discharge power for configurations 1 and 2 and (b) percentage increase in turbine power than 2-stage configuration 1

Cost of turbomachinery (compressor and turbine) per kW of discharge power is shown in Figure 4. The turbomachinery cost decreases with increase in the discharge power (refer to Table 2). Therefore, the cost trends are opposite to the turbine discharge power shown in Figure 3. There is no significant difference in the cost of turbomachinery between configurations 1 and 2.

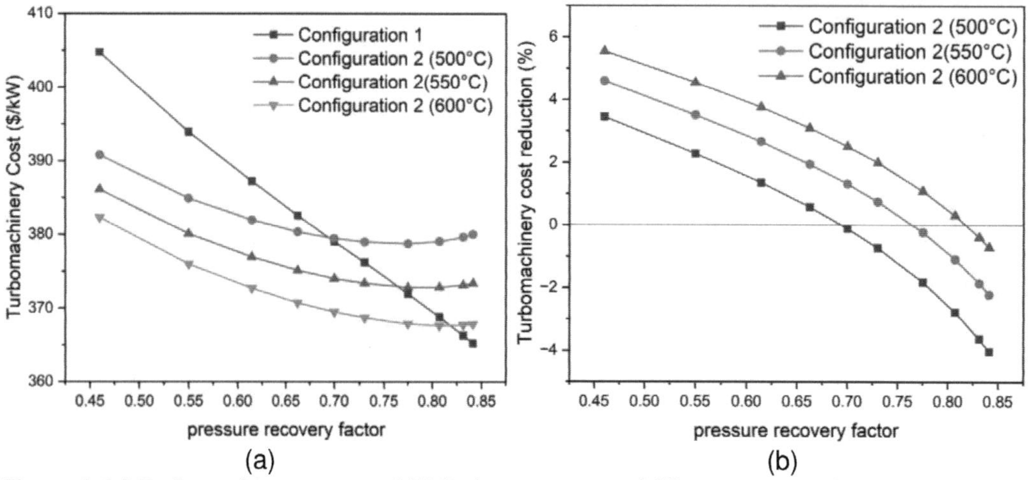

Figure 4. (a) Turbomachinery cost per kW discharge power and (b) percentage reduction in cost for configuration 2 than 2-stage configuration 1

The cost of PBTES system for configurations 1 and 2 per kW of discharge power is plotted in Figure 5. The cost of packed bed constitutes two components: TES material cost and pressure vessel cost. For the given mass flow rate and charging time, the mass of the TES material is constant irrespective of operating pressure (refer to Eq. 7). However, the mass of the pressure vessel increases with increase in the operating pressure. The cost of PBTES per kW of power delivered is lower for configuration 2 than configuration 1. Figure 5b shows the percentage reduction in the cost of PBTES with configuration 2. Configuration 2 showed higher cost reductions for a lower pressure recovery factor due to the delivery of high power. At higher pressure recovery factors, the reduced RTE and turbine power caused less reductions in the PBTES cost. At δ_p = 0.614 (43-70 bar), the percentage reduction is 51, 54, and 56%, respectively for 500, 550, and 600°C.

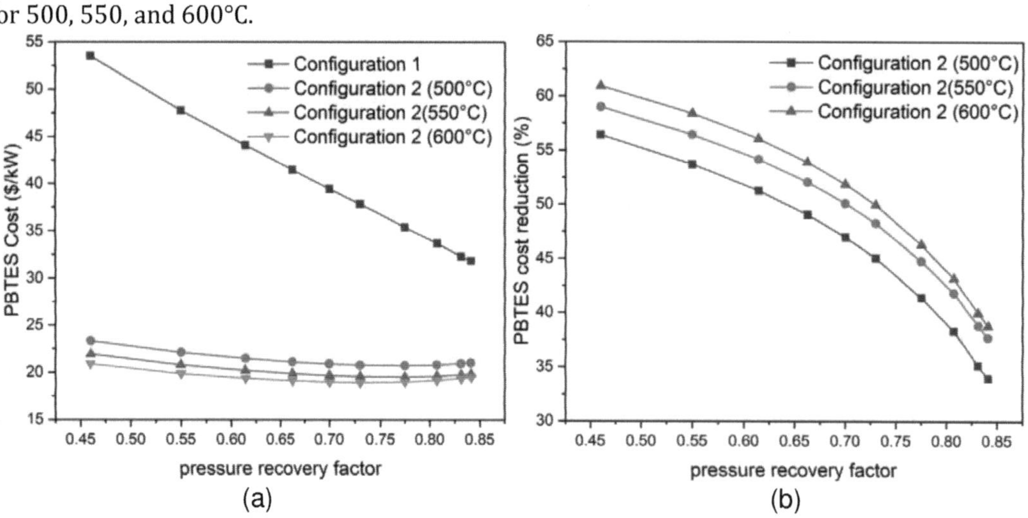

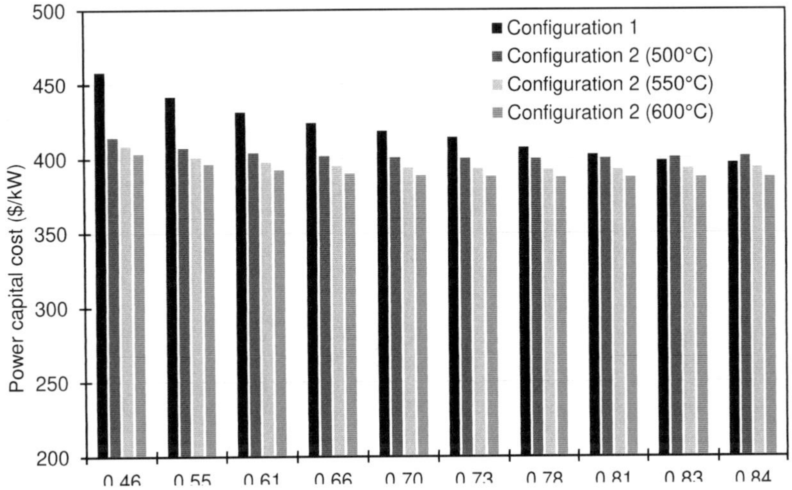

(c)

Figure 5. (a) Cost of PBTES per kW discharge power, (b) percentage reduction in PBTES cost and (c) total power capital cost of configurations 1 and 2

Power capital cost is shown in Figure 5c for configurations 1 and 2. We have observed that the cost of PBTES in configuration 2 is significantly reduced, but no significant reduction in turbomachinery cost. Since the magnitude of the turbomachinery cost is higher than the PBTES cost, the power capital cost does not significantly decrease with the help of configuration 2. The power capital cost is reduced by 6, 8, and 9% at δ_p value of 0.614 for compressor temperature limits of 500, 550, and 600°C, respectively. Overall, the proposed Tri-ACES configuration showed an enhanced RTE, and slightly reduced costs compared to the conventional ACAES system.

4.2 Heat extracted for space/ district heating during charging

The outlet temperature of air and heat output of compressor 3 is shown in Fig. 6c. It is observed that the compressor 3 outlet temperature decreases with the increase in the maximum outlet temperature limit of the 1st and 2nd compressor. This is attributed to the decreased compression ratio of the 3rd compressor with increase in T_{limit}. The heat recovery temperatures are in the range of 68-190°C up to δ_p of 0.7 (maximum RTE and turbine power zone), which are suitable for district heating application. Further, compressor 3 is estimated to consume 8-26% of the total input power for different T_{limit} values up to δ_p of 0.7.

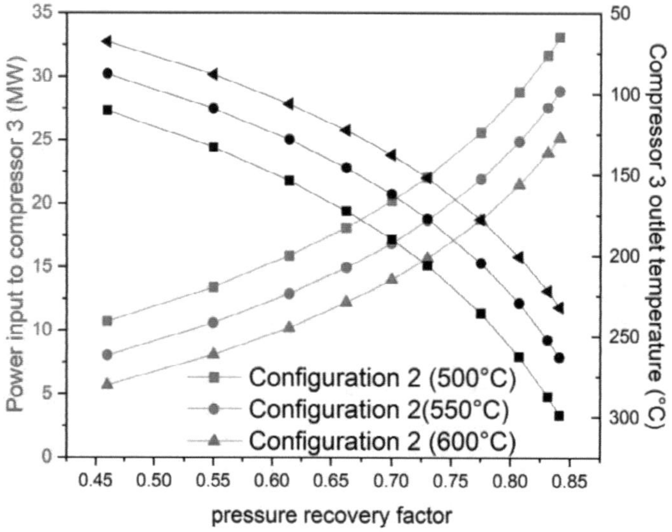

Figure 6. Power input/ heat extracted and outlet temperature of compressor 3 (configuration 2)

A preliminary analysis was conducted to determine the significance of heat extracted from the CAES heat exchanger for district heating. The schematic of the CAES heat exchanger is shown in Figure 7. The hot air from the compressor 3 is used to heat the return water from district heating network. The number of UK houses that can be fed with the extracted heat is calculated and shown in Table 5. For the analysis, the following assumptions are considered.

- Heat load of a UK home is considered as 2646.25 kWh/household/year.
- Local community is located at 10 km distance from CAES facility.
- CAES operates at 365 cycles a year of 10 hours per run.
- Return and delivery temperatures of water are considered as 60 and 95°C, respectively.

Two operating pressure ranges are considered for the analysis, which are close to Huntorf and McIntosh CAES plants (Table 5). Results indicated that the Tri-ACAES system during the charging process can provide heat for up to 16,390 UK homes located within 10 km distance if connected with an existing district heating network.

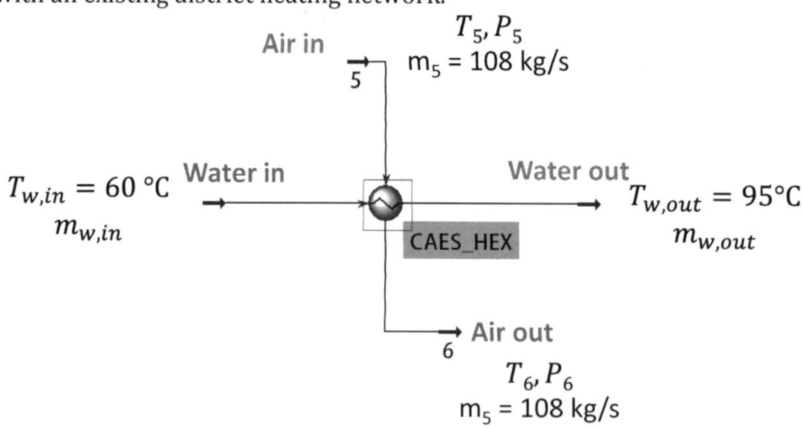

Figure 7. Schematic of CAES heat exchanger for district heating application

Table 5 Heat load calculations for district heating application

Pressure range (bar)	δ_p	Compressor outlet temperature, T5 (°C)			Water mass flow rate (kg/s)			Number of UK homes		
		500°C	550°C	600°C	500°C	550°C	600°C	500°C	550°C	600°C
43-70	0.614	153.7	128.4	106.1	63.74	44.58	27.63	13400	9371	5808
53-80	0.663	172.5	145.9	122.6	77.95	57.84	40.18	16390	12160	8445

4.3 Heating and cooling mode during idle time of CAES

The ARC subsystem shown in Figure 1b is operated to generate heating and cooling when the excess renewable energy is too low to operate the entire compression train, and/ or the air reservoir is in fully charged condition. The generated heat and cold energy are stored and dispatched as per the demand. The cooling temperatures and cooling power of the Tri-ACAES system during heating and cooling mode is presented in Figure 8. The cooling power is calculated by considering $T_{1'}$ as 15°C. The Tri-CAES system delivers cooling temperatures up to -98°C. The cooling power produced is 50% of the heat output from the compressor (averaged over entire operating pressure range).

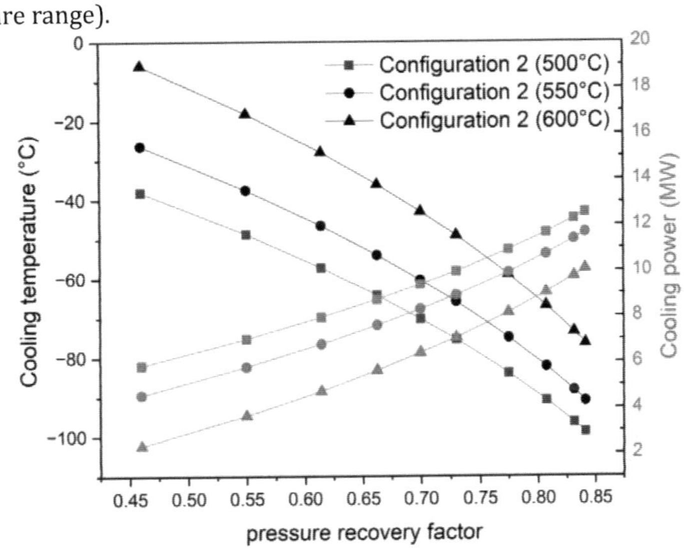

Figure 8. Cooling power and cooling temperatures of Tri-ACAES system

5. Conclusions

In the present study, a new configuration of ACAES system is proposed and compared with conventional ACAES system. The following are the major outcomes from the thermodynamic and economic analysis.

- The proposed Tri-ACAES system enhances the RTE and power capacity up to 5.8% and 33%, respectively than conventional ACAES system.
- The total cost of the system is reduced by 10% and the cost of PBTES is reduced by up to 60% with the Tri-ACAES system.
- The proposed Tri-ACAES system is capable of supplying heat to 16,390 UK homes situated within 10 km of CAES facility every day.
- It also has the potential to generate cooling temperatures up to -98°C when the excess renewable energy is too low to operate the ACAES system.

If a CAES system does not engage in the energy market daily, it's payback period will be prolonged due to its significant initial investment. However, the proposed Tri-ACAES system will enhance the operability of the CAES system and offer diverse participation opportunities such as heating, cooling, and power. These opportunities can help reduce the payback period to some extent. However, integrating such a Tri-ACAES system into real-world scenarios is challenging by the coexistence of heating, cooling, and electricity demands near the CAES facility and balancing these diverse energy needs efficiently.

Nomenclature

Symbol	*Description*	*Unit*
C	cost	$
C_p	specific heat	J/kg-K
D_b	diameter of packed bed	m
H	hight	m
M	mass	kg
$\dot{m}$	mass flow rate	kg/s
Nu	Nusselt number	
P	pressure	Pa
r_c	compression ratio	
r_t	expansion ratio	
S	allowable stress	Pa
t	thickness	m
T	temperature	K
V	volume	m^3
Greek symbols		
η	efficiency	
ρ	density	kg/m^3
γ	specific heat ratio	
β	Mass overrating factor	
Acronyms		
CAES	compressed air energy storage	
TES	thermal energy storage	
PB	packed bed	
AR	aspect ratio	
HEX	heat exchanger	
LMTD	log mean temperature difference	
RTE	round-trip efficiency	
OD	Outer diameter	

Subscripts	
amb	ambient
b	bed
ch	charging
comp	compressor
disch	discharging
in	inlet
isen	isentropic
max	maximum
out	outlet
p	polytropic
turb	turbine

Acknowledgement

The authors would like to thank EPSRC Hi-CAES project (EP/W027372/1) for the funding support.

References

1. Bazdar E, Sameti M, Nasiri F, Haghighat F. Compressed air energy storage in integrated energy systems: A review. Renew Sustain Energy Rev [Internet]. 2022;167(February):112701. Available from: https://doi.org/10.1016/j.rser.2022.112701
2. Zhang X, Li Y, Gao Z, Chen S, Xu Y, Chen H. Overview of dynamic operation strategies for advanced compressed air energy storage. J Energy Storage [Internet]. 2023;66(March):107408. Available from: https://doi.org/10.1016/j.est.2023.107408
3. King M, Jain A, Bhakar R, Mathur J, Wang J. Overview of current compressed air energy storage projects and analysis of the potential underground storage capacity in India and the UK. Renew Sustain Energy Rev [Internet]. 2021;139(December 2020):110705. Available from: https://doi.org/10.1016/j.rser.2021.110705
4. Fajinmi O, Munda JL, Hamam Y, Popoola O. Compressed Air Energy Storage as a Battery Energy Storage System for Various Application Domains: A Review. Energies. 2023;16(18).
5. Nageswara Rao M. Materials for Gas Turbines – An Overview. In: Advances in Gas Turbine Technology. Intech; 2011.
6. Cárdenas B, Garvey S. A directly charged thermal store for compressed air energy storage systems. J Energy Storage. 2023;71(July).
7. Sciacovelli A, Li Y, Chen H, Wu Y, Wang J, Garvey S, et al. Dynamic simulation of Adiabatic Compressed Air Energy Storage (A-CAES) plant with integrated thermal storage – Link between components performance and plant performance. Appl Energy [Internet]. 2017;185:16–28. Available from: http://dx.doi.org/10.1016/j.apenergy.2016.10.058
8. Kotowicz J, Job M, Brzęczek M, Nawrat K, Mędrych J. The methodology of the gas turbine efficiency calculation. Arch Thermodyn. 2016;37(4):19–35.
9. RETORET JP, MOREAU M. Reciprocating Compressors. Vol. 40, Ind. Petrole. Eur. 1972. 69–75 p.
10. Meng H, Wang M, Olumayegun O, Luo X, Liu X. Process design, operation and economic evaluation of compressed air energy storage (CAES) for wind power through modelling and simulation. Renew Energy [Internet]. 2019;136:923–36. Available from: https://doi.org/10.1016/j.renene.2019.01.043
11. Shuailing L, Guoyuan M, Shuxue X, Yuexuan G, Xiaoya J, Guoqiang W. A review of reverse Brayton air cycle refrigerators. Int J Refrig [Internet]. 2023;150:200–14. Available from: https://doi.org/10.1016/j.ijrefrig.2023.02.016
12. Wilson DG. The Design of High-Efficiency Turbomachinery and Gas Turbines. MIT Press. 2014. 1–5 p.
13. Mersch M, Sapin P, Olympios A V., Ding Y, Mac Dowell N, Markides CN. A unified framework for the thermo-economic optimisation of compressed-air energy storage systems with solid and liquid thermal stores. Energy Convers Manag [Internet]. 2023;287(April):117061. Available from: https://doi.org/10.1016/j.enconman.2023.117061
14. Luyben WL. Capital cost of compressors for conceptual design. Chem Eng Process - Process Intensif [Internet]. 2018;126(January):206–9. Available from: https://doi.org/10.1016/j.cep.2018.01.020
15. Nabat MH, Zeynalian M, Razmi AR, Arabkoohsar A, Soltani M. Energy, exergy, and economic analyses of an innovative energy storage system; liquid air energy storage (LAES) combined with high-temperature thermal energy storage (HTES). Energy Convers Manag [Internet]. 2020;226(May):113486. Available from: https://doi.org/10.1016/j.enconman.2020.113486
16. Saleh Kandezi M, Mousavi Naeenian SM. Thermodynamic and economic analysis of a novel combination of the heliostat solar field with compressed air energy storage (CAES); a case study at San Francisco, USA. J Energy Storage. 2022;49(August 2021):104111.
17. Stack DC, Curtis D, Forsberg C. Performance of firebrick resistance-heated energy storage for industrial heat applications and round-trip electricity storage. Appl Energy [Internet]. 2019;242(October 2018):782–96. Available from: https://doi.org/10.1016/j.apenergy.2019.03.100
18. Crotogino F, Mohmeyer KU, Scharf R. Huntorf CAES: More than 20 Years of Successful Operation. Solut Min Res Inst Spring Meet. 2001;(April):351–7.

Journal of Physics: Conference Series **2929** (2024) 012004 doi:10.1088/1742-6596/2929/1/012004

Modelling the hydraulic performance of an array of centrifugal pumps operating under a variable pressure in a hydro-pneumatic energy storage system

Andrea Cassar[1], Tonio Sant[1*] and Andrew Borg[1]

[1] Department of Mechanical Engineering, Faculty of Engineering, University of Malta, Msida, Malta
*E-mail: tonio.sant@um.edu.mt

Abstract. The large pressure variation in hydro-pneumatic energy storage systems presents a major challenge for maintaining a high operating efficiency of the turbomachinery used in the hydro-electric energy conversion processes. This paper applies numerical modelling to compare the hydraulic performance of two pumping unit variants used for charging an offshore hydro-pneumatic energy storage (HPES) accumulator. The first unit consists of a single large centrifugal pump operated at variable speed while the second pumping unit is made up of an array of smaller, dynamically similar fixed-speed pumps. A number of series/parallel configurations for the latter option were derived in order to optimise the operational efficiency at various states of charge of a gas-charged accumulator, specifically between 80 and 200 bar. A comparative analysis of the operational hydraulic efficiencies for both pumping units shows that using an array of small fixed-speed pumps arranged in series/parallel configurations is more energy efficient than using a single large pump operated at variable speed.

Nomenclature

g	Acceleration due to gravity, (m²/s)
n	Number of pump impellors for a multi-stage pump
D_2	Impeller diameter of new pump, (m)
H_2	Head across new pump at best efficiency point (BEP), (m)
$H_{2(required)}$	Required head across new pump, (m)
$H_{2(new)}$	New head across smaller pump, (m)
P	Nominal power of large pump, (MW)
P_{in}	Input or shaft power, (MW)
P_{out}	Output or hydraulic power, (MW)
Δp_{P1-P3}	Pressure difference across pumps 1-3, (bar)
Δp_{P4-P6}	Pressure difference across pumps 4-6, (bar)
$\Delta p_{2(required)}$	Required pressure difference across new pump, (bar)
Q_2	Flow rate across new pump at BEP, (m³/s)
Q_{P1-P3}	Flow rate across pumps 1-3, (m³/s)
Q_{P1-P3}	Flow rate across pumps 4-6, (m³/s)
r_v	Volumetric ratio, (-)
ρ	Density of seawater, (kg/m³)
ω	Rotational speed of impeller, (rad/s)

1. Introduction

The renewable energy sector has seen an increase in the use of a variety of energy storage systems (ESSs) addressing the supply-demand mismatch and grid instability problems resulting from the intermittent nature of renewable energy sources such as wind and solar power [1, 2]. Generally, energy can be stored in a variety of forms, including mechanical, chemical, thermochemical, electrical and thermal [3]. Recently there has been a growing interest to deploy ESSs offshore [4, 5]. A majority of offshore ESS technologies involve compressed air energy storage (CAES). An example is underwater compressed air energy storage (UWCAES). The pressure containment may consist of closed underwater ballons made out of a flexible material, as for example in the Energy Bags concept, that retains the compressed air at the hydrostatic pressure [6, 7]. The containment may also consist of a rigid material, such as subsea concrete reservoirs. These are also able to maintain a fixed pressure equal to the hydrostatic pressure by having an opening to allow sea water to flow in and out of the reservoirs, forming a liquid piston. One example is the BAROMAR concept [8]. UWCAES concepts are able to maintain high efficiency given the air compressors and expanders work at a constant pressure load, that is equal to the hydrostatic pressure. However, such concepts will have to be deployed in deep waters (typically sea depths >500m) to be able to attain a high energy storage density in terms of kilowatt-hour per cubic metre.

Hydro-pneumatic energy storage (HPES) is an alternative form of CAES technology that is also being developed for the offshore environment [9]. A HPES system may be assumed to be made up of two main subsystems: a pressure containment system (PCS) which is pre-charged with compressed air and an energy conversion unit (ECU) that incorporates a coarse filter, a pump or array of pumps, an energy recovery system and an inverter-generator, as shown in Figure 1. The ECU powers the pump/s in periods of abundant renewable energy supply to inject seawater into the PCS, which in-turn compresses the air inside the vessel and stores energy. During periods of high energy demand, the ECU allows the pressurised seawater inside the PCS to expand through

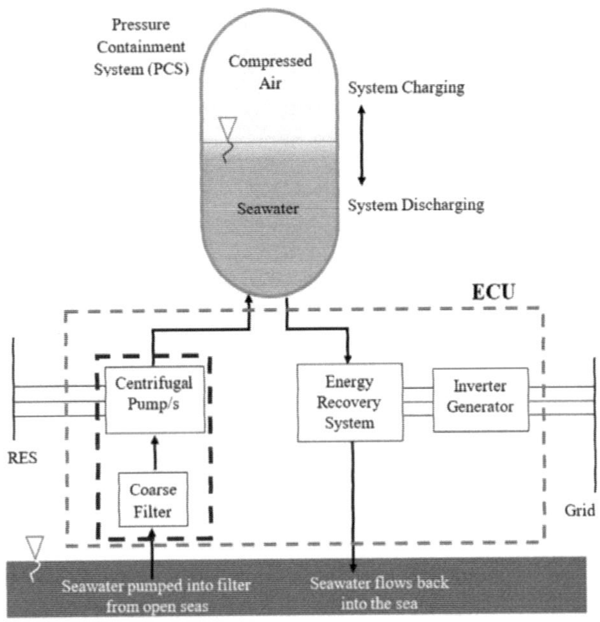

Figure 1. Schematic diagram showing the working principle of a HPES system.

an energy recovery system (such as a turbine) to generate electricity. Unlike UWCAES, PCS pre-charging allows HPES systems to be deployed in shallow waters, in the close vicinity of offshore wind farms. Furthermore, deploying the PCS in a subsea environment enables the system to exploit the excellent heat sink/heat source characteristics of sea water, thus maintaining quasi-isothermal conditions. In fact, a high thermal efficiency (>93%) has already been achieved through an offshore-based HPES prototype deployed in the Grand Harbour in Malta [10]. Although HPES is a straightforward technology, the pumping unit will be exposed to a wide operating pressure window, thereby having an impact on its operational efficiency. The main scope of this paper is to investigate the feasibility of replacing the single large multi-stage centrifugal pump operated at variable speed within the ECU with an array of smaller pumps with geometrically similar impellers operating at fixed speed, arranged in different series/parallel configurations depending on the pressure within the PCS. The paper uses numerical modelling to compare the hydraulic efficiency of these two options. The motivation for the latter option is driven by the potential cost reduction opportunities arising from the avoidance of a variable speed drive and use of smaller centrifugal pumps that are more widely available in the offshore industry. Furthermore, for conditions where the ECU is operating well below its overall rated power, using smaller pumps enables some individual electric drives to operate at a power level that is closer to the rated valve, thus enabling them to work at a higher electrical efficiency. This paper is organised as follows: Section 2 presents the approach for deriving the hydraulic performance curves for the small fixed speed pumps from the larger centrifugal pump. Section 3 presents the numerical models for simulating the hydraulic performance of the pump/s supplying water to the PCS during HPES charging. Section 4 discusses the results comparing the efficiency for the two different pumping unit options.

2. Derivation of Small Fixed-Speed Pump Characteristics

The hydraulic performance characteristics of a large (Megawatt-scale) eight-stage centrifugal pump were used to derive the performance curves for a single stage impeller by dividing the head and input power values by eight. The characteristics for a single impellor of this Megawatt-scale pump are presented in Figures 2 – 4. The following dimensionless parameters were then used to derive the characteristics of a geometrically similar impeller for the small pump:

$$Head\ coefficient, C_H = \frac{gH}{\omega^2 D^2} \quad (1)$$

$$Discharge\ coefficient, C_Q = \frac{Q}{\omega D^3} \quad (2)$$

$$Power\ coefficient, C_P = \frac{P}{\rho \omega^3 D^5} \quad (3)$$

Moreover, the pump efficiency η_p can be also expressed in terms of the dimensionless coefficients:

$$\eta_p = \frac{C_H C_Q}{C_P} \quad (4)$$

Ignoring the effects of impeller surface roughness ratio ε/D, as well as the Reynolds number effects, which could be ignored for highly turbulent flows [11], Eqts. (1) – (4) could be applied to

Journal of Physics: Conference Series **2929** (2024) 012004 doi:10.1088/1742-6596/2929/1/012004

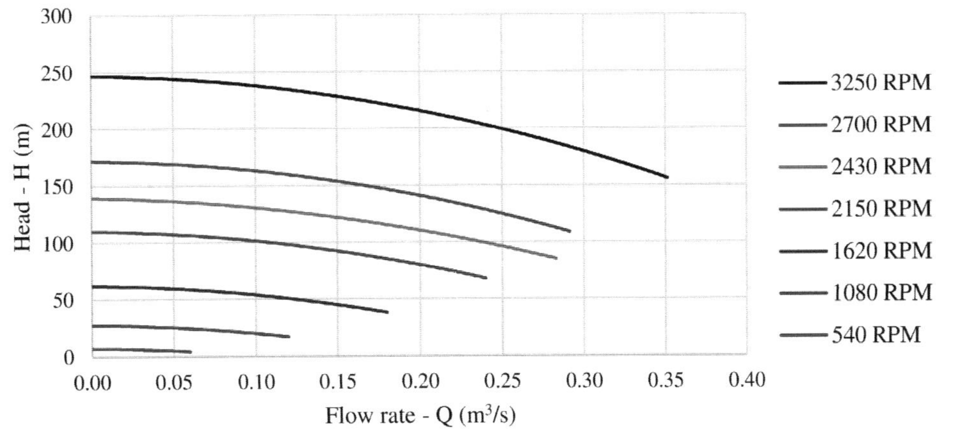

Figure 2. Head (m) against Flow rate (m^3/s) for a single stage of the MW-scale pump.

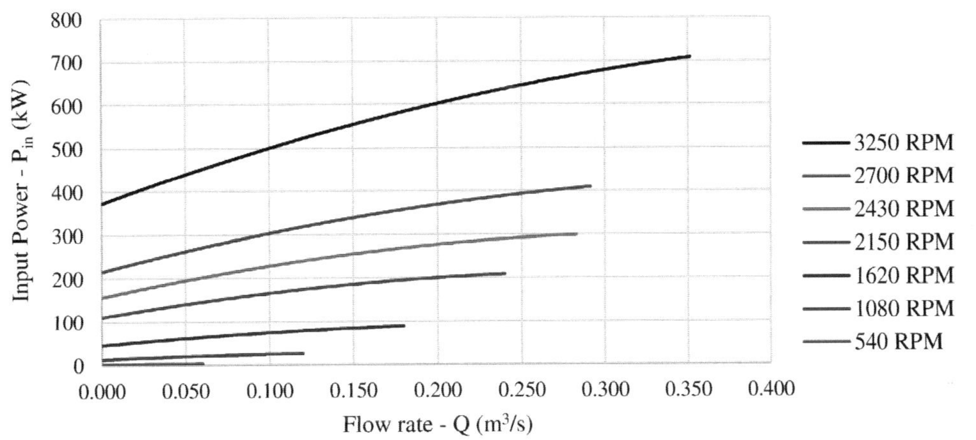

Figure 3. Input Power (kW) against Flow rate (m^3/s) for a single stage of the MW-scale pump.

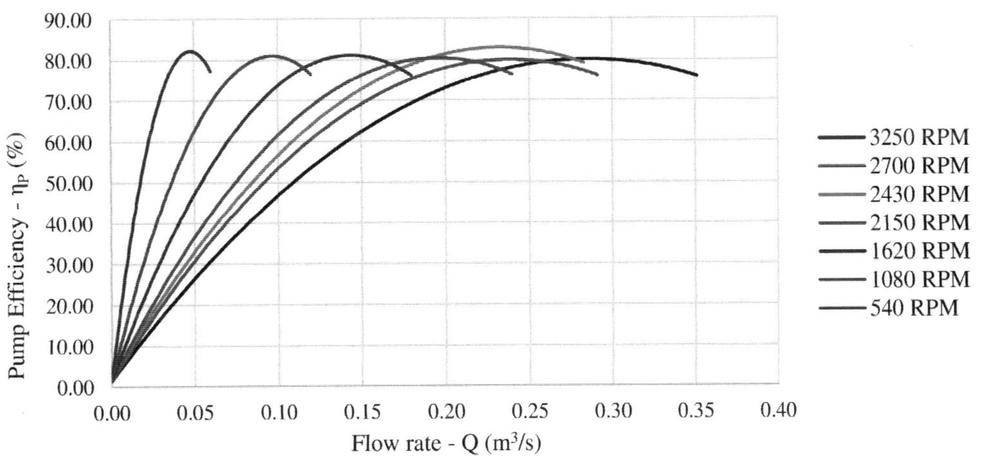

Figure 4. Pump Efficiency (%) against Flow rate (m^3/s) for a single stage of the MW-scale pump.

Journal of Physics: Conference Series **2929** (2024) 012004 doi:10.1088/1742-6596/2929/1/012004

find the dimensionless pump parameters, including the efficiency, at the different speeds of the impeller of the large MW-scale pump. The best efficiency point (BEP) at each pump speed was identified and the corresponding dimensionless parameters were noted as C_Q^*, C_H^*, C_P^* and η_P^*. Since the dimensionless parameters given by (1) - (3) for both impellers are now equal, the following similarity rules could be used to find the volume flow rate, differential head and input power of the small pump impeller.

$$\frac{Q_2}{Q_1} = \frac{\omega_2}{\omega_1}\left(\frac{D_2}{D_1}\right)^3 \tag{5}$$

$$\frac{H_2}{H_1} = \left(\frac{\omega_2}{\omega_1}\right)^2 \left(\frac{D_2}{D_1}\right)^2 \tag{6}$$

$$\frac{P_{in,2}}{P_{in,1}} = \left(\frac{\rho_2}{\rho_1}\right)\left(\frac{\omega_2}{\omega_1}\right)^3 \left(\frac{D_2}{D_1}\right)^5 \tag{7}$$

where suffixes *1* and *2* denote the impellers of the large and small pumps, respectively.

According to Cutajar et al. [12], the optimal pressure ratio that yields the highest storage capacity per unit of steel mass used for the HPES accumulator was found to be around 2.5. The peak working pressure for the accumulator was set to 200 Bar, which is a typical working pressure for subsea pressure containments currently used in the offshore industry. Thus, for an optimal pressure ratio of 2.5, the initial charging pressure was set to 80 Bar. To determine an appropriate size for the small pump, it was first assumed that six identical pumps consisting of 3 pumps in series connected to another 3 pumps in series, as shown in Figure 5, would meet the mid-point pressure of 140 Bar. The discharge for this configuration is given by (8):

$$Q_{P1-P3} = \frac{1}{2}\left(\frac{P}{\Delta p_{P1-P3}}\right) = Q_{P4-P6} = Q_2 \tag{8}$$

For the sake of clarity, the variables used in the relation above are also being indicated in Figure 5. The discharge value in the above relation was divided by two since the two sets of three identical pumps each are connected in parallel. It was decided that the small pump should operate at a fixed-speed (ω_2) equal to the electricity supply grid frequency, i.e., 50 Hz. The impeller diameter for the small pump D_2 was derived using Eqt. (2) as follows:

$$D_2 = \left(\frac{Q_2}{C_Q^* \omega_2}\right)^{\frac{1}{3}} \tag{9}$$

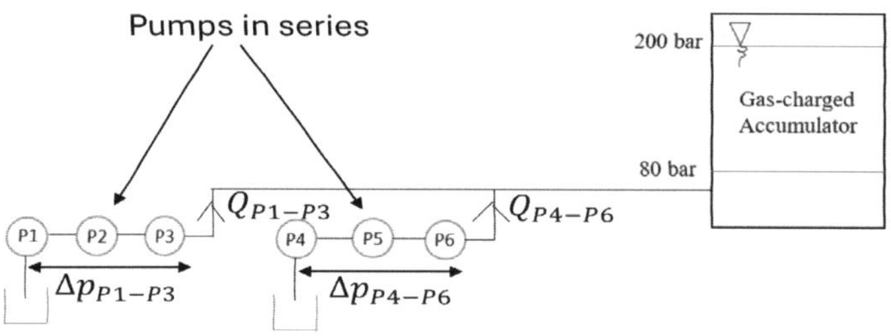

Figure 5. A simple schematic diagram consisting of six fixed-speed small pumps connected to the accumulator to find the volume flow rate and pressure difference across the new small pump.

To derive the pressure rise across a single small pump, the pressure difference Δp_{P1-P3} had to be divided by three (Figure 5). Thus, the required pressure difference at design conditions, $\Delta p_{2(required)}$, across the new small pump is given by (10):

$$\Delta p_{2(required)} = \frac{\Delta p_{P1-P3}}{3} \tag{10}$$

$H_{2(required)}$ could be obtained by dividing the value above by ρg. Furthermore, the actual head for a single impellor of the small pump, given by H_2, was calculated using equation (6). Since H_2 was observed to be less than the required overall head of the small pump ($H_{2(required)}$), a number of impellers had to be added to the new multi-stage pump to make sure that the required head at the BEP is reached. The number of impellers required for the small pump n was found using (11) as follows:

$$n = \frac{H_{2(required)}}{H_2} \tag{11}$$

The value of n was rounded to the next larger whole number. Moreover, the similarity laws given by (5) - (7) were then used to calculate the new heads and input powers for various flow rate values. Although the efficiencies for the larger and small pump impellors should in theory be equal for complete dynamic similarity, pumps with flow regimes having larger Reynolds numbers and lower roughness ratios tend to have higher efficiencies, as explained by White [13]. This phenomenon is known as efficiency scale-up [14]. According to Peltz et al. [15], the ratio of inefficiencies from a larger pump to a smaller pump impellor for a hydraulically smooth surface is given by equation (12):

$$\frac{1-\eta_2}{1-\eta_1} = \left(\frac{Re_1}{Re_2}\right)^{\alpha} \tag{12}$$

Where η_1 and η_2 along with Re_1 and Re_2 are the efficiencies and Reynolds numbers for the large and small pump impellors, respectively. The constant α can be approximated to 0.1 as proposed in [14]. Once the head, input power and efficiency were calculated at various flow rate values, the characteristic curves for the new (small) multi-stage pump were obtained. Table 1 shows a comparison of the operating data between the large MW-scale pump and the new small fixed-speed pump with geometrically similar impellors.

Table 1. A comparison of the design data for the larger and smaller pumps.

Parameter	Centrifugal Pump 1	Centrifugal Pump 2
Impeller Diameter (m)	0.38	0.31
Number of Stages	8	3
Pump speed (rpm)	Variable Speed: 540 - 3250	Fixed-speed: 3000
Pressure range (Bar)	≤200 at 3250 rpm	27 - 43
Capacities (m^3/s)	≤0.351	≤0.190
Efficiency at BEP (%)	82.95	82.24

3. Numerical Modelling

3.1 Numerical model for pumping unit charging an HPES accumulator

Numerical models in MATLAB® were used to simulate the performance of the centrifugal pumps during the charging cycle of a gas-charged accumulator. The assumed volume V_2 of the accumulator operating between 80 and 200 Bar was 4080 m^3, resulting in an energy storage capacity of 7.6 MWh. This capacity was computed considering an ideal isothermal HPES system operating subsea in a hydrostatic pressure p_{hyd} with a volumetric ratio of 2.5, following the relation below:

$$W_{isothermal} = p_{max}V_2(\ln(r_v)) - p_{hyd}V_2(r_v - 1) \quad (13)$$

The development of the numerical models involved the use of several Simulink® and Simscape™ blocks which were parametrised using the values found in the previous calculations. Since the Simulink® modelling of an electrical variable speed drive system for the single large pump would have gone beyond the scope of this study, a number of models were constructed, each having the same centrifugal pump operated at a different speed.

The main blocks used for the modelling of both pumping units include the *hydraulic fluid, solver configuration, centrifugal pump, hydraulic pipeline, check valve, constant head tank* and *gas-charged accumulator*. The centrifugal pump characteristics was parametrised by inputting the previously derived pump characteristic curves in the form of the pressure differential, brake power and flow rate vectors. Moreover, the hydraulic pipelines enabled the flow of fluid from seawater to the pump and from the pump to the accumulator. A check valve was inserted in order to ensure no reverse flow of seawater in the pipeline. In fact, the maximum passage area for the check valve was set as the area of the pipe. Apart from the blocks mentioned above, pressure, flow rate and rotational power sensors were connected to the system. This enabled the numerical model to measure the total pressure, flow rate and input power across any pump at specific head values. Figure 6 shows the complete Simulink® model for the single large pump operating at 2430 rpm. The filter pressure drop was modelled using the two-dimensional model of Sun *et al* [16].

The following two separate numerical modelling approaches were adopted for modelling the pump – accumulator interactions, with the latter operating between a pressure range of 80 and 200 Bar:

(1) Constant head approach: in this case the 80 to 200 Bar range was divided into six equally spaced pressure bands (PBs), i.e., between 80 and 100 bar, between 100 and 120 bar etc. Simulations were performed on the mid-point value of each PB; this implies that the minimum accumulator's pressure value was that of 90 bar. This approach facilitated the readings of the pressure, flow rate and input power at specific head values. Given that the minimum accumulator pressure is 90 bar, it was observed that only speeds between 2430 and 3250 rpm for the single large pump could reach such pressure value.

(2) Variable head approach: this involved the use of the gas-charged accumulator model which simulates the increase in head at the pump with time. Simulations were conducted for a specific stop time which will be explained in the following section 4

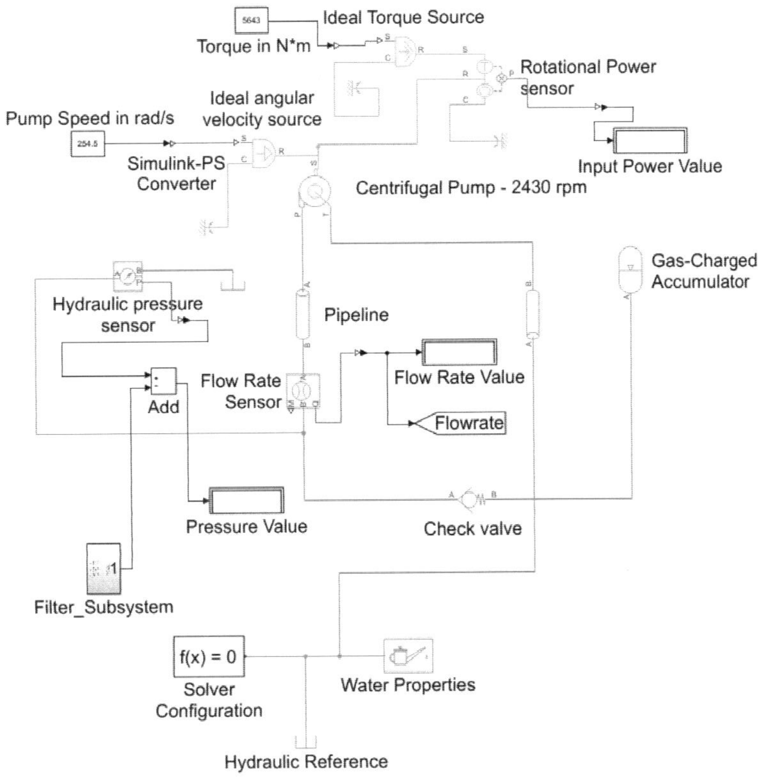

Figure 6. A screenshot of the complete Simulink® model for the large pump operating at 2430 rpm.

3.2 Deriving feasible array configurations for the small pumps operating at fixed speed

The same model used to simulate the single large pump operated with a VSD (section 3.1) was modified to simulate a pumping unit consisting of an array of smaller identical fixed-speed pumps. It was decided to limit the number of smaller pumps to six since a large number of pumps would be impractical to manage and these would produce a lower overall efficiency due to Reynolds number effects. Moreover, these pumps can be arranged in a number of configurations depending on the required overall differential head and flow rate of the system. A total of 17 theoretical pump configurations were derived. Some of these configurations include pumps arranged in series, in parallel or a combination of both. These are shown in Figure 7. However, following preliminary calculations, some configurations were found to be unfeasible given that they could not reach the minimum pre-charge pressure of 90 Bar of the HPES accumulator. Only 6 out of the 17 configurations were retained. These include configurations *C4*, *C6*, *C10*, *C12*, *C14* and *C16*. The selection process is outlined in Figure 7.

4. Results and Discussion

This section presents the results comparing the hydraulic performance of the single large pump operated at variable speed with that of the six small pump configurations (*C4*, *C6*, *C10*, *C12*, *C14* and *C16)*. The results are presented for the constant head and variable head numerical approaches highlighted in section 3.1. The Simulink® models outputted the pressure and flowrate across the pumps, together with the input shaft power. The hydraulic power output and

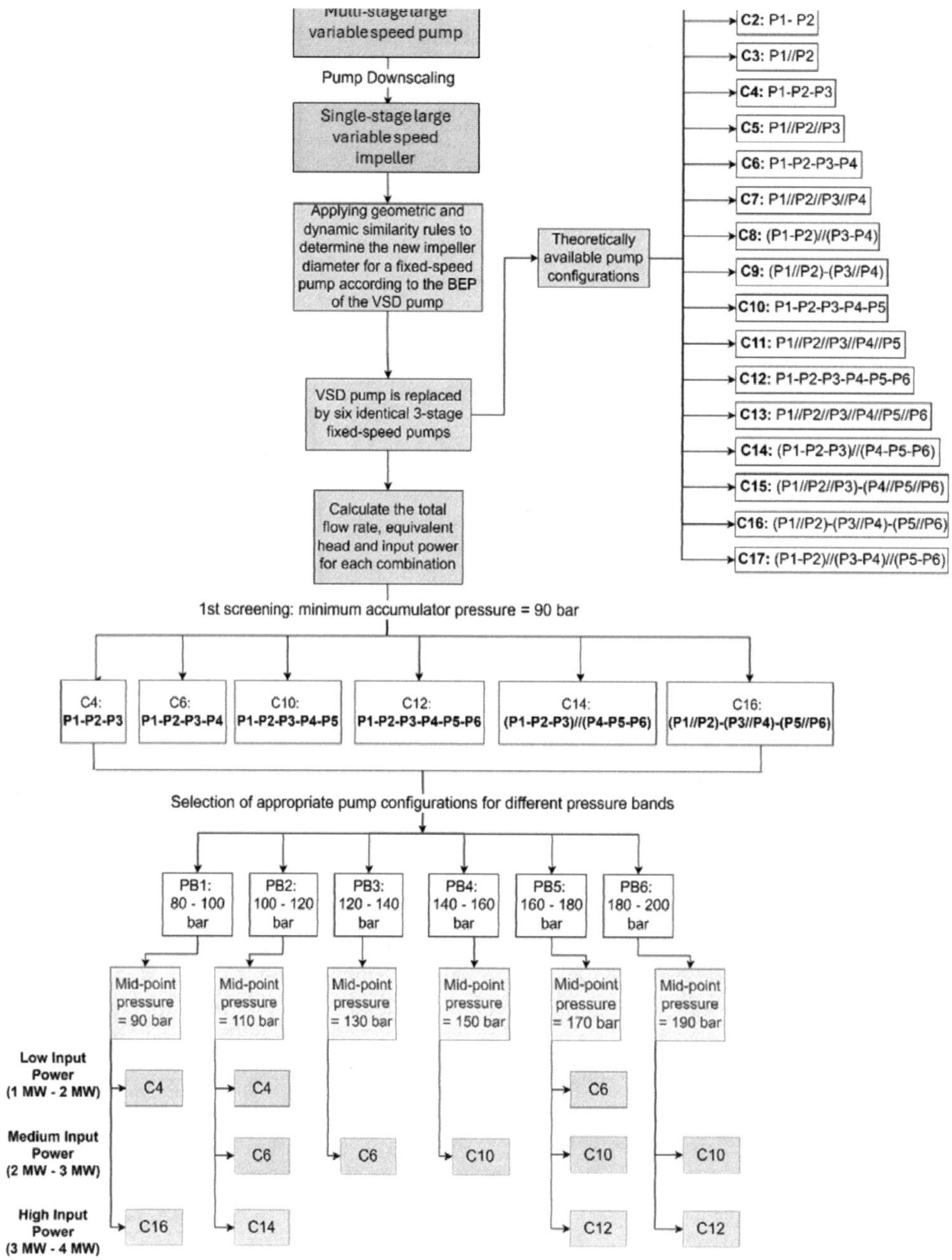

Figure 7. A flowchart showing the decision-making process for selecting the feasible pump configurations. Symbols '//' and '-' denote pumps in parallel and series, respectively

efficiency were computed separately. The hydraulic power output of each pump was calculated by multiplying the flowrate by the pressure differential across the pump. The hydraulic efficiency was found by dividing the sum of the hydraulic out power values for all pumps by the sum of the power inputs.

4.1 Numerical results for a Constant Head Approach

Table 2 presents the maximum efficiencies and corresponding speed and input power values for the single large MW-scale pump using the constant head approach. The pump yielded a maximum pump efficiency of 77.77% at 3050 rpm for a constant head tank of 130 bar while the lowest efficiency (45.35%) was achieved at 3250 rpm for a constant head tank of 190 bar. The maximum efficiency values for the small pumps operating at fixed speed are shown in Table 3. It should be noted that in this case the overall pump array efficiency is presented. Pump configuration C12 (six identical pumps arranged in series) yielded a maximum pump array efficiency of 81.51% for a constant head tank of 170 bar. On the other hand, a minimum efficiency of 79.38% was achieved by pump configuration C6 (four identical pumps arranged in series) for a constant head tank of 130 bar. Table 3 presents the results for the other constant head tank

Table 2. Maximum pump efficiencies obtained for a single large pump operated at variable speed.

Accumulator Pressure (bar)	Pump Speed (rpm)	Pump Efficiency (%)	Input Power (MW)
90	2600	77.50	2.82
110	2750	77.37	3.27
130	3050	77.77	4.50
150	3250	77.41	5.39
170	3250	72.14	5.02
190	3250	45.35	3.93

Table 3. Maximum overall pump array efficiencies obtained for smaller fixed-speed pumps.

Accumulator Pressure (bar)	Pump Configuration	Pump Array Efficiency (%)	Input Power (MW)
90	C16	80.67	3.57
110	C6	81.26	2.40
130	C6	79.38	2.33
150	C10	80.54	2.98
170	C12	81.51	3.60
190	C12	80.16	3.52

values. Pump configurations C10 and C16 represent five pumps arranged in series and three sets of two parallel pumps connected in series with each other, respectively. When comparing the results obtained for both pumping units, it is evident that the small fixed-speed pump configurations yielded higher overall pumping unit efficiencies than the single large variable-speed pump. Additionally, it can be observed that the efficiencies for the small pump configurations remain quite high along the accumulator pressure range whereas the efficiency for the single large pump starts decreasing after reaching 150 bar. Moreover, the input power values for the single large pump were observed to be higher than those for the combination of pumps. This is due to the fact that the fixed-speed pumps are smaller in size compared to the single MW-scale pump, thus achieving a lower overall input power.

4.2 Numerical results for a Variable Head Approach

The constant head tank was then replaced by a gas-charged accumulator having a pre-charged pressure of 80 bar. In this case, the charging time for the accumulator was set to around 4.4 hours (16,000 s) by changing the stop time of the Simulink® simulation. The latter was deemed appropriate in order for the accumulator to fully charge. Figure 8 below shows a typical Simscape™ plot for the variation of pressure inside the accumulator with charging time. The graph shown below is specifically for pump configuration *C10* (configuration explained in Fig. 7). Similar graphs were obtained for the other configurations.

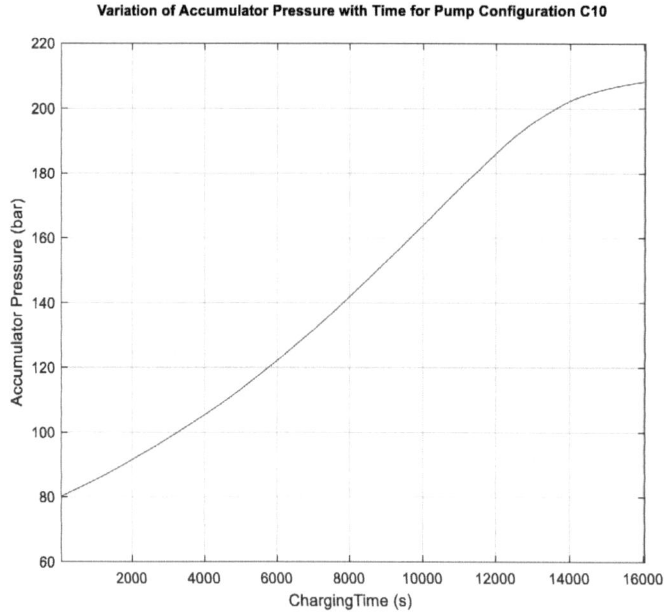

Figure 8. A graph of accumulator pressure (bar) against charging time (s) for pump configuration C10.

Plots similar to that in Figure 8 were extracted for each pump configuration. A number of points having the charging time in common were then selected along the curves and the pressure values were multiplied by the flow rates to find the output power. The pump efficiencies were finally calculated by dividing the output powers by the input powers which were obtained from the characteristic curves. These values were tabulated in Excel® and the variation of the efficiency

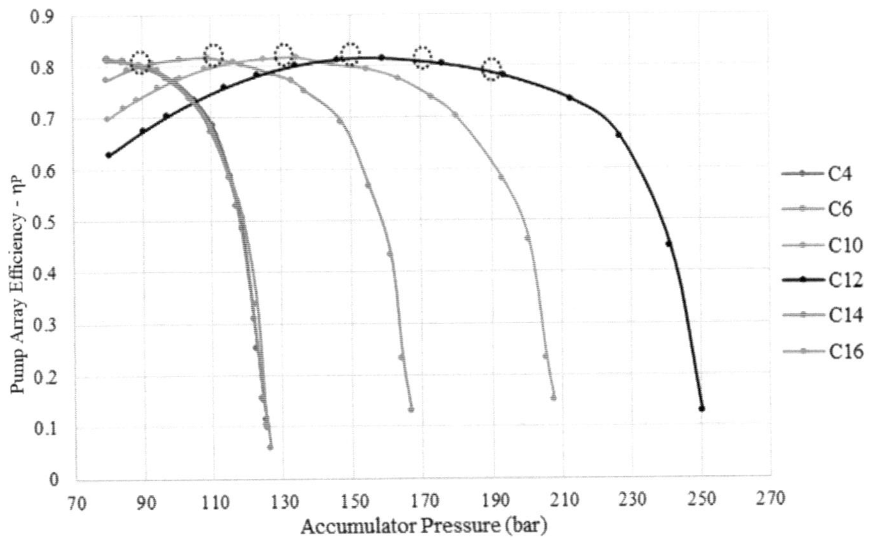

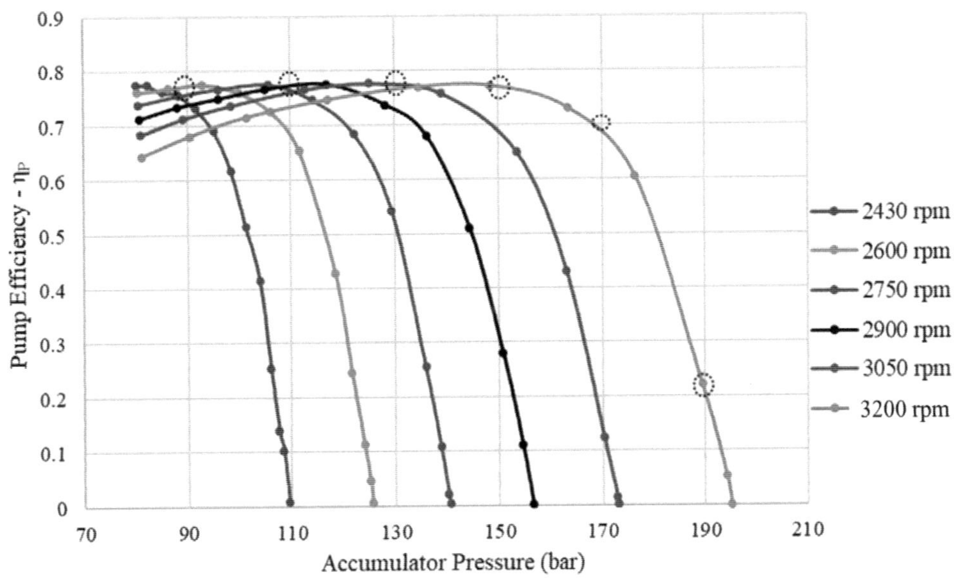

Figure 9. Graph of pump efficiency against accumulator pressure for the six feasible pump configurations operating at constant speed (top) and for the single large pump operated at variable speed (bottom).

with accumulator pressure was plotted for both pumping units, as shown in Figure 9. It can be observed that each pump configuration can achieve a high efficiency over a small range of accumulator pressures; this justifies the need for different pump configurations in order to charge the accumulator. The dotted circles in both figures represent the best efficiency point (BEP) at the mid-point value of each pressure band (PB). Furthermore, the graph represented by Figure 9 confirms that the various small pump configurations yield higher overall efficiencies throughout the charging phase of the accumulator as compared to the large pump operated at variable speed.

5. Conclusions and Future Work

The work carried out throughout this study was based on the development of two numerical models which simulate the operating characteristics of large-scale centrifugal pumps for HPES applications. The numerical analysis compared the performance of a single large pump operated at variable speed with an equivalent pumping unit consisting of a multitude of dynamically similar but smaller pumps, each operating at constant speed. The following list comprises the main findings from the present study:

- The numerical model of the single MW-scale centrifugal pump operated at variable speed yielded a maximum operating efficiency of 77.77% at 3050 rpm, for an input power value of 4.5 MW and constant head tank of 1293 metres (130 bar).
- A maximum operating efficiency of 81.51% was obtained for pump configuration C12, which is made up of six identical centrifugal pumps arranged in series and operated at 3000 rpm. The input power and constant head tank values at this efficiency were found to be 3.6 MW and 1691 metres (170 bar), respectively.
- Overall, results show that an array of constant-speed pumps with adjustable interconnections for series/parallel flow configurations is more energy efficient than a single large pump operated with variable speed. The former option could result in a more economical solution for charging hydro-pneumatic energy storage systems due to the fact that (1) smaller pumps are cheaper to buy given that they are produced in larger volumes, and (2) a variable speed drive (VSD) is not required.

One main limitation of arranging several fixed-speed pumps in series and parallel configurations without the use of a VSD is the lack of control over the input power to the system. Although the results presented in this study show that the maximum operating efficiencies for an array of fixed-speed centrifugal pumps are greater than those achieved with a single large centrifugal pump operated with a VSD, there are several factors that demand further research and analysis.

Future work will involve a detailed analysis to evaluate the reductions in the capital and operating costs that could be potentially achieved when replacing the single variable-speed pump with multiple fixed-speed pumps. Moreover, the possibility of employing a different type of pump, such as positive displacement pumps, which are known to retain a high efficiency over a wider pressure range, can also be considered in future work.

6. References

[1] Rahman MM, Oni AO, Gemechu E, Kumar A. Assessment of energy storage technologies: A review. *Energy Conversion and Management.* 2020; 223: 113295, doi.org/10.1016/j.enconman.2020.113295.
[2] Koohi-Fayegh S and Rosen MA. A review of energy storage types, applications and recent developments. *Journal of Energy Storage.* 2020; 27: 101047, doi.org/10.1016/j.est.2019.101047.
[3] Guney MS and Tepe Y. Classification and assessment of energy storage systems. *Renewable and Sustainable Energy Reviews.* 2017; 75: 1187-1197, doi.org/10.1016/j.rser.2016.11.102
[4] Arellano-Prieto, Y, Chavez-Panduro E, Salvo Rossi P, Finotti, F. Energy Storage Solutions for Offshore Applications. *Energies.* 2022: 15, 6153. https://doi.org/10.3390/en15176153
[5] Yu H, Yang X, Chen H, Lou S, Lin Y. Energy Storage Capacity Planning Method for Improving Offshore Wind Power Consumption. *Sustainability.* 2022; 14(21):14589. doi.org/10.3390/su142114589
[6] Pimm AJ, Garvey SD and De Jong M. Design and testing of Energy Bags for underwater compressed air energy storage. *Energy.* 2014; 66: 496-508. doi.org/10.1016/j.energy.2013.12.010.

[7] Wang H, Wang Z, Liang C, Carriveau R, Ting DS-K, Li P, Cen H, Xiong W. Underwater Compressed Gas Energy Storage (UWCGES): Current Status, Challenges, and Future Perspectives. *Applied Sciences*. 2022; 12(18):9361. doi.org/10.3390/app12189361

[8] "Technology," [Online]. Available: https://www.baro-mar.com/. [Accessed 8 May 2024].

[9] Silva J, Maia Alves J, Sensitivity analysis of levelized cost of hydro-pneumatic electricity storage for grid support applications. *Renewable Energy*. 2023; 219, Part 1: 119285. doi.org/10.1016/j.renene.2023.119285.

[10] Buhagiar D, Sant T, Farrugia RN, Aquilina L, Farrugia D and Strati FM. Small-scale Experimental Testing of a Novel Marine Floating Platform with Integrated Hydro-pneumatic Energy Storage. *Journal of Energy Storage*. 2019; 24:1-6. doi.org/10.1016/j.est.2019.100774.

[11] Cengel YA and Cimbala JM. Fluid Mechanics: Fundamentals and Applications, New York: McGraw-Hill, 2006.

[12] Cutajar C, Sant T, R. Farrugia RN and Buhagiar D. A Software Tool for the Design and Operational Analysis of Pressure Vessels used in Offshore Hydro-pneumatic Energy Storage. *Journal of Energy Storage*. 2021; 40: 1-9. doi.org/10.1016/j.est.2021.102750.

[13] White FM. Fluid Mechanics, 8th Edition, New York: McGraw-Hill Education, 2016.

[14] "Efficiency scale-up," [Online]. Available: https://www.ksb.com/en-global/centrifugal-pump-lexicon/article/efficiency-scale-up-1116642. [Accessed 5 February 2022].

[15] Pelz PF and Stonjek SS. Introduction of an Universal Scale-Up Method for the Efficiency of Axial and Centrifugal Fans. in *ASME Turbo Expo 2014: Turbine Technical Conference and Exposition*, 2014.

[16] Sun H, Bu S and Luan Y. A high-precision method for calculating the pressure drop across wire mesh filters. *Chemical Engineering Science*. 2015;127: 143-150. doi.org/10.1016/j.ces.2015.01.023.

Journal of Physics: Conference Series **2929** (2024) 012005 doi:10.1088/1742-6596/2929/1/012005

Comparing the Pressure Containment Requirements of Three Subsea Hydro-Pneumatic Energy Storage Concepts

Tonio Sant[1*], Luke Aquilina[1], Charise Cutajar[1], Daniel Buhagiar[2] and Daniel Baldacchino[2]

[1] Department of Mechanical Engineering, Faculty of Engineering, University of Malta, Msida, Malta
[2] FLASC B.V., Paardenmarkt 1, 2611 PA Delft, The Netherlands

*E-mail: tonio.sant@um.edu.mt

Abstract. Hydro-pneumatic energy storage has significant potential for being deployed subsea and co-located with offshore wind farms. The pressure containment storing the compressed air is the bulkiest and most expensive component of such a storage technology. This paper evaluates three different subsea hydro-pneumatic energy storage concepts, comparing their volumetric storage densities, the steel and concrete anchoring requirements and cost of the pressure containment. The pressure variation experienced by the pressure containment across the storage cycle for the three variants, which would impact the efficiency of energy conversion machinery, is also compared. It is shown that having separate accumulators, each designed for a different operating pressure, offers the potential of reducing the cost of the pressure containment by up to 40% as compared to that for the simple concept having a single accumulator. A novel approach is presented to achieve such cost reduction without constraining the energy conversion machinery to operate over a wider operating pressure range.

1. Introduction

There is a growing interest today to develop mechanical energy storage technologies for the offshore environment which can provide long duration storage of energy originating from large scale offshore wind farms [1, 2]. Co-locating energy storage within the same wind farm footprint results in more efficient use of marine space, while also avoiding land requirements to house storage infrastructure. Some research activities in offshore energy storage explored the potential of using existing geological features below the seabed (e.g. saline aquifers) for storing compressed air. These are however geographically restricted [3]. Others focused on marinising existing advanced-adiabatic compressed air energy storage (AA-CAES) technology by deploying the system on a floating platform, with the requirement to accommodate the thermal energy storage (TES) system [4]. Other offshore storage concepts are based on exploiting the body of sea water for its hydro-static pressure using seabed-mounted infrastructure. These can be mainly classified in three categories: (1) Subsea pumped-hydro energy storage (PHES) [5, 6], (2) underwater compressed air energy storage (UWCAES) [7, 8], and (3) Buoyancy/gravity-based energy storage [9]. A major barrier for such technologies is the large sea depth required (typically >500m) to achieve reasonable storage densities. This presents significant challenges for co-locating such storage infrastructure with wind farms as the latter are not envisaged to be developed on a commercial scale in waters deeper than 350 m within the coming decade.

Hydro-pneumatic energy storage (HPES) is an alternative form of compressed air energy storage (CAES) technology, in which liquid pistons driven by hydraulic machinery are used to compress and expand the gas. HPES can be considered as a pumped-hydro system, with the hydrostatic head created through air pressurization [10-13]. When deployed offshore, HPES systems can utilise the immense thermal mass of the seawater to achieve near-isothermal compression and expansion. It has already been demonstrated by the authors that a high thermal efficiency (>93%) is achievable when an HPES system is deployed in sea water [14]. This avoids the need for the TES system normally required for AA-CAES systems [15], hence facilitating the deployment of CAES offshore. Furthermore, unlike other mechanical energy storage technologies being developed to be deployed at sea, offshore HPES does not rely on hydro-static pressure. This offers significant potential for integrating storage within the same wind farm marine space.

Utilising closed pressure containment systems (PCSs) pre-charged with the compressed gas avoids the need for air from the open atmosphere to operate HPES systems. This enables such systems to be deployed as a fully subsea solution, without the need for offshore platforms and associated interconnections (e.g. power cables and hoses) to accommodate hardware (e.g. hydraulic machinery) above the sea surface. However, the pressure containment remains the bulkiest and most expensive component of HPES systems. The main objective of this paper is to evaluate three different subsea HPES concepts, comparing the volumetric storage densities, the steel and concrete anchoring requirements and the indicative cost of their PCS. The paper is organized as follows: Section 2 describes the three HPES concepts under consideration while Section 3 presents the underlying theory forming the basis of this study. Sections 4 and 5 describe the numerical modelling carried out together with the presentation and discussion of key results.

2. Subsea HPES Concepts

Figure 1 presents the three subsea HPES concepts being investigated in the present study. It is assumed that the PCS is built from offshore pipeline sections, which are the cheapest industrialised solution available for storing gases offshore. The first concept (Fig. 1a) is the simplest form of HPES concept (HPES-1P), consisting of a single high pressure vessel (HPV) that acts as a liquid piston accumulator. Initially, the vessel is pre-charged with compressed air to a pressure p_1 which is higher than the hydrostatic pressure p_H. During periods of high renewable energy system (RES) availability, a high-pressure pump-turbine (HPT) is operated in pump mode to inject sea water into the pressure vessel, thus compressing the air and storing the energy. During periods of low-RES production, the gas is expanded by allowing the pressurised sea water to flow through the pump-turbine which is operated in turbine mode to generate electricity.

A disadvantage of the HPES-1P concept is the inefficient use of steel for compressed air as the entire pressure containment must be designed for the peak rated pressure p_2, including the proportion of volume occupied by the water at the fully charged state (Fig. 1a). The second HPES concept (Fig. 1b) aims to mitigate this problem by using two separate pressure vessels (HPES-2P), designed for different peak operating pressures, with the low-pressure vessel (LPV) acting as a liquid piston accumulator with a peak pressure p_L, while the high-pressure vessel (HPV) is only used to store compressed air from p_1 up to a peak working pressure p_2. To store energy, an air compressing-expanding machine (AEM) is operated as a compressor to transfer compressed air from the LPV into the HPV. The AEM may be a conventional piston-type compressor-motor or centrifugal-type turbomachine, or a liquid piston compressor-expander to achieve higher thermal

Journal of Physics: Conference Series **2929** (2024) 012005		doi:10.1088/1742-6596/2929/1/012005

efficiencies [16]. A low-pressure pump-turbine (LPT) is operated in tandem with the AEM in pump mode to inject sea water into the LPV to maintain a constant pressure p_L in this vessel. This prevents the pressure difference across the AEM from increasing significantly during the charging process as air is transferred from the LPV to the HPV. Given that, for a given sea depth H, the external hydrostatic pressure p_H is fixed, the pressure difference across the LPT will also be constant, independent of the state of charge, hence allowing this hydraulic machine to be operated at optimal efficiency. During the discharging process, the AEM is operated as an expander to generate electricity, allowing the high-pressure gas in the HPV to flow back into the LPV. The LPT is operated in turbine mode to generate electricity, while maintaining a fixed pressure difference across its as the water is ejected from the LPV into the open sea.

While the HPES-2P concept reduces the overall steel required for pressure containment, the fact that p_L is larger than p_H constrains the AEM to operate over a wider pressure range than in the case of the HPT of the HPES-1P concept (the ratio $\Delta p_2/\Delta p_1$ in Fig. 1b than that in Fig. 1a). This

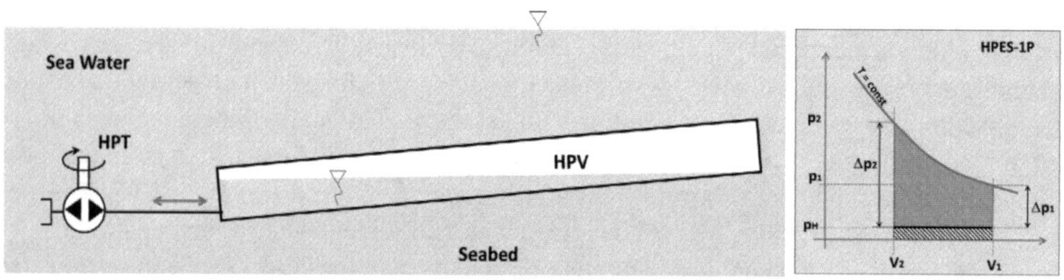

(a) HPES-1P

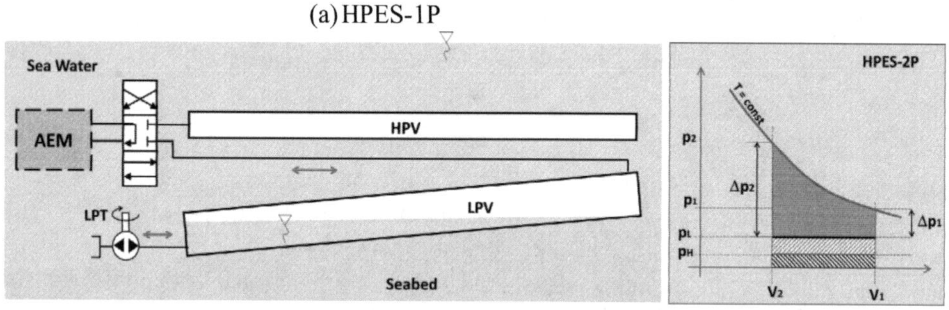

(b) HPES-2P

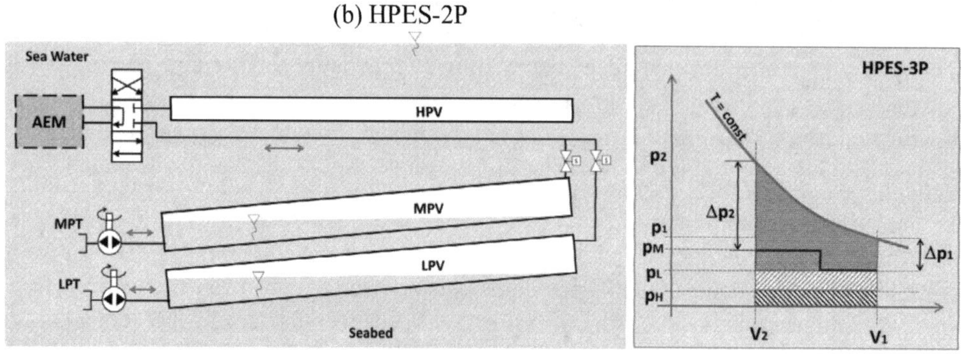

(c) HPES-3P

Figure 1. The 3 different subsea HPES concepts (left) and the corresponding pressure-volume charts (right) assuming an isothermal air compression/expansion

Journal of Physics: Conference Series **2929** (2024) 012005 doi:10.1088/1742-6596/2929/1/012005

may thus lead to a lower efficiency in the energy conversion. To mitigate this problem a third HPES concept (Fig 1c) with three pressure vessels (HPES-3P) is being proposed. It operates in a similar way to the HPES-2P concept, with each vessel designed for a different working pressure range. Two pressure vessels act as liquid piston accumulators: a low-pressure vessel (LPV) operating at a fixed pressure p_L and a medium pressure vessel (MPV) operating at a fixed pressure p_M, such that $p_M > p_L$. Both accumulators are interconnected to separate pump-turbines, corresponding to the low (LPT) and medium (MPT) pressure values. During the initial part of the charging cycle, only the AEM and LPT are operated. Once the LPV is full of water, the LPT is switched off and the MPT takes over. Given that $p_M > p_L$, the maximum pressure difference across the AEM, Δp_2, is reduced.

3. Theoretical Background

3.1 Models for the Ideal Storage Capacity

Model for HPES-1P system (Fig. 1a): Consider a simple closed HPES system with a single pressure vessel consisting of a pipeline of diameter D and length L located at a sea depth H and undergoing isothermal compression/expansion between the pre-charge pressure p_1 and a peak operating pressure p_2. Assuming a perfect gas, it can be shown that the ideal storage capacity is equal to:

$$E = p_2 \, V \frac{\ln(r_p)}{r_p} - \rho g V H \left(1 - \frac{1}{r_p}\right) \tag{1}$$

where r_p is the pressure ratio equal to p_2/p_1, p_1 being the initial pre-charge pressure and V equal to the volume of the pressure vessel; g is the acceleration due to gravity.

Model for HPES-2P system (Fig. 1b): For the HPES system with one liquid piston accumulator and one air-only vessel, it can be shown that, for purely isothermal conditions, the ideal energy storage capacity is equal to

$$E = p_2 \, V \frac{\ln(r_p)}{r_p} + p_L \, V_L \ln\left(\frac{p_2}{p_L}\right) - \rho g V_L H \tag{2}$$

In this case, V and V_L are the volumes of the HPV and LPV, respectively. The pressure ratio r_p is defined as p_2/p_1, where p_1 and p_2 are the initial pre-charge and peak operating pressures for the air-only HPV vessel. The LPV is pre-charged to p_L. The system will be fully charged when the LPV is full of sea water, i.e. all air transferred to the HPV. The operational pressures for the system to operate should satisfy the following:

$$p_2 > p_1 > p_L > \rho g H$$

Applying fundamental principles of mass conversation, it can be shown that the diameter D' of the LPV is equal to

$$D' = D\sqrt{\left(\frac{p_2 - p_1}{p_L}\right)} \tag{3}$$

where D is the diameter of the HPV.

Model for HPES-3P system (Fig. 1c): For the HPES system with two liquid piston accumulators and one air-only vessel, the ideal energy storage for isothermal conditions is determined from:

$$E = p_2 \, V \frac{\ln(r_p)}{r_p} + p_L \, V_L \ln\left(\frac{p_2}{p_L}\right) + p_M \, V_M \ln\left(\frac{p_2}{p_M}\right) - \rho g H (V_L + V_M) \tag{4}$$

r_p and p_2 are defined as for the HPES-2P system. The HPV is initially pre-charged up to pressure p_1, with pressure ratio r_p defined as p_2/p_1. The MPV and LPV are pre-charged to p_M and p_L, respectively. The system will be fully charged when all compressed air is transferred to the HPV, i.e. when both the MPV and LPV are full of sea water. The operational pressures for the system to operate should satisfy the following:

$$p_2 > p_1 > p_M > p_L > \rho g H$$

From simple mass conservation considerations and isothermal processes, it can be shown that the diameters D, D' and D'' of the HPV, LPV and MPV, respectively, are related in accordance with the following two relations:

$$D' = D''\left(\frac{p_1+p_2}{2p_L}\right)^{1/4} \tag{5}$$

$$D'' = D\left[\left(\frac{p_2-p_1}{2}\right)\left(\frac{2}{p_L(p_1+p_2)}\right)^{1/2}\right]^{1/2} \tag{6}$$

The above relations are derived on the assumption that the masses of the pre-charge air in the MPV and LPV are the same and that the intermediate pressure p_M is equal to

$$p_m = \sqrt{p_L\left(\frac{p_1+p_2}{2}\right)} \tag{7}$$

3.2 Pressure difference factor

As already explained earlier, the pressure difference across the LPT and MPT (Figs 1b and c) is kept constant. However the pressure difference across the HPT (Fig. 1a) and AEM (Fig. 1b,c) machines varies across the charging/discharging cycle. To quantify this variation, which in reality impacts the operating efficiency of such machines, we define the pressure difference factor as the ratio of the pressure difference across the HPT and AEM when the HPES system is fully charged divided by that when the system is fully discharged ($\Delta p_2/\Delta p_1$, see Fig. 1):

$$f_{rp} = \frac{p_2-p_k}{p_1-p_r} \tag{8}$$

p_k and p_r are both equal to the hydrostatic pressure p_H ($=\rho g H$) for the HPES-1P concept and equal to p_L for the HPES-2P concept. In the case of the HPES-3P concept, p_k and p_r are equal to p_M and p_L, respectively. An f_{rp} value equal to unity would represent an isobaric energy process.

3.3 Model for Steel Requirements for Pressure Vessels

As already discussed in section 2, the pressure vessels for all of the HPES concepts are assumed to consist of offshore pipeline sections. The wall thickness t for a given pipeline section is determined using a model which combines the thin cylinder theory with the von Mises relation:

$$t = \frac{\sqrt{3}}{4} f \frac{D}{k_w}\left(\frac{p_i-\rho g H}{\sigma_y}\right) + c \tag{9}$$

where D is the diameter of the pressure cylinder composed of steel having material yield strength σ_y, p_i is the internal peak design working pressure, f is the factor of safety, k_w is the welded joint factor and c is the added corrosion allowance. Given that for pipelines the diameter is significantly smaller than the length, we can approximate the ends to flat discs instead of hemi-spherical shells when computing the steel mass requirements.

3.4 Model for Concrete Requirements for Pressure Vessels

Concrete is added to the steel pipelines for anchoring purposes, to ensure that the PCS is sinkable and has a gravitational load that is large enough to retain a stable position on the seabed. We define the buoyancy factor *BF* factor as the ratio of the total mass of the sea water displaced by the

Journal of Physics: Conference Series **2929** (2024) 012005 doi:10.1088/1742-6596/2929/1/012005

pressure containment system volume (steel plus concrete), divided by the total system mass. Based on Archemedes' principle, BF equal to unity represents a pressure containment system that is neutrally buoyant. To have a sinkable system, BF <1.

3.5 Cost for Pressure Containment
The overall cost for the pressure containment per kWh, excluding installation, is computed from

$$C_{tot} = S_{st}c_{st} + S_{crt}c_{crt} \tag{10}$$

Where C_{tot} is the overall specific cost (Euro per kWh), S_{st} and S_{crt} are the mass of steel and concrete required (tonne per kWh), while c_{st} and c_{crt} are the specific costs of steel and concrete (Euro per tonne).

4. Methodology

A parametric analysis was carried out to compare the PCS steel and concrete requirements of three subsea HPES concepts (Fig. 1) based on the theoretical equations presented in Section 3. The analysis was organized into three case studies, as shown in Table 1, to study the dependency on the pressure ratio r_p, peak operating pressure p_2 and pre-charge pressure p_L of the LPV, while maintaining the design parameters in Table 2 fixed. Case studies 1 and 3 assumed fixed values for p_2 equal to 200 Bar. This is a typical design working pressure for steel pipelines currently used in the offshore energy industry. Case studies 2 and 3 assumed an r_p value equal to 2.5, which from recent studies by the authors has shown to yield the lowest steel mass per unit of storage capacity for the HPES-1P concept [17]. In each case study, the following parameters were used as basis for comparing the three HPES concepts:

1. Specific volume (S_v), defined as the volume of the steel pressure vessels per unit of storage capacity (m^3/kWh)
2. Specific steel (S_{st}), defined as the mass of steel pressure vessels per unit of storage capacity (tonne/kWh)
3. Specific concrete (S_{crt}), the mass of anchoring concrete per unit of storage capacity (tonne/kWh)
4. Pressure difference factor (f_{rp}), (Eqt 8)
5. Overall specific cost (C_{tot}), (Eqt. 10)

It should be noted that comparing the thermal efficiency of the three concepts to gauge the ability in achieving near isothermal behaviour, through the exchange of heat between the compressed air and the surrounding sea water, was not within the scope of this study.

Table 1. Three case studies modelled

Case Study	Variable	Variable Range	Other fixed parameters
1	Pressure ratio (r_p)	$1.4 < r_p < 3.4$	p_2= 200 Bar p_L= 25 Bar H= 30 m
2	Peak pressure (p_2)	$125 < p_2 < 300$ Bar	r_p = 2.5 p_L= 25 Bar H= 30 m
3	Pre-charge pressure (p_L)	$5 < p_L < 35$ Bar	p_2 = 200 Bar; r_p = 2.5 H= 30 m

Table 2. HPES parameters retained fixed throughout parametric analysis

Parameter	Value	Parameter	Value
D	0.5 m	k_w	0.9
L	600 m	c	3.0 mm
H	30 m	BF	≤0.9
σ_y	483 N/mm^2	c_{st}	1000 Eur/tonne
f	1.7	c_{crt}	122 Eur/tonne

5. Results and Discussion

5.1 Case Study 1: Influence of Pressure Ratio

Figures 2 (a) – (f) show the influence of the operating pressure ratio (r_p=p_2/p_1) on the specific volume (S_v), the specific steel and concrete requirements (S_{st} and S_{crt}), and the overall specific cost (C_{tot}), in terms of Euro/kWh. In all three concepts, the lowest volumetric requirements are reached at around 2.3<r_p<2.6, above which volume per unit of kWh starts to increase (Fig. 2a). The HPES-1P concept has a larger volumetric density as compared to the HPES-2P and HPES-3P concepts due to the lower specific volume S_v. It is to be noted that S_v considers only the volume requirements for the steel pressure vessels and does not include the volume of the concrete ballast. As observed from Fig. 2b, the HPES-2P concept has the lowest specific steel requirements, with the specific steel S_{st} being less than half than that of the HPES-1P concept. On the other hand, the HPES-2P concept has larger concrete requirements than for the HPES-3P concept, which increase with the pressure ratio (Fig. 2c). The HPES-1P concept does not require concrete to achieve the maximum allowable BF value of 0.9 (Table 2).

The overall PCS costs per kWh as determined using Eqt. 10 are compared in Fig. 2d. The minimum cost for the HPES-1P concept is reached at around r_p=2.6, beyond which it starts to increase again. The costs for the other two concepts reduce gradually with r_p, at the expense that higher r_p may constrain the AEM to work outside optimal design conditions for a longer period of time. The percentage cost reduction achieved by the HPES-2P/3P concepts as compared to the simple HPES-1P one is shown to increase with r_p (Fig. 2e). The percentage cost reduction is significant, reaching 35 to 40% at r_p~2.6, beyond which further potential reductions are small.

Fig. 2(f) plots the pressure difference factor (f_{rp}, Eqt. 8) with the pressure ratio. Higher values for r_p cause the HPT (in the HPES-1P concept) and the AEM (in the HPES-2P/3P concepts) to operate over a wider pressure range across the full charging/discharging cycle due to the larger f_{rp} values, thus making it more challenging to retain a high efficiency for the energy conversion process. It is observed that while the HPES-2P has the overall lowest cost for pressure containment (Fig. 2d), it encounters the highest f_{rp} values. One the other hand, the HPES-3P concept retains lower f_{rp} values, with the value at r_p=2.5 being approximately equal to that for the HPES-2P concept.

5.2 Case Study 2: Influence of Peak Operating Pressure

Figures 3 (a) – (e) show the dependence on the peak design pressure p_2, when keeping r_p and p_L fixed at 2.5 and 25 Bar, respectively. The benefits of operating at higher peak pressures are evident: (1) higher volumetric storage densities (lower S_v values, Fig. 3a); (2) reduced mass of

Journal of Physics: Conference Series **2929** (2024) 012005 doi:10.1088/1742-6596/2929/1/012005

steel and concrete (Figs. 3b and c), leading to higher achievable cost reductions for the HPES-2P/3P concepts with respect to the HPES-1P concept (Fig. 3d), and (3) reduced operating pressure range for the AEM as demonstrated by the lower f_{rp} values (Fig. 3e).

5.3 Case Study 3: Influence of Pre-Charge Pressure

Figures 4(a)-(f) show the influence of the pre-charge pressure p_L of the HPES-2P and HPES-3P low pressure vessel (LPV, see figs 1b,c) for r_p and p_2 fixed at 2.5, 200 Bar, respectively. The results for the HPES-1P system are also included for comparative purposes only, as p_L is only relevant to the other two concepts. The pre-charge pressure for the HPES-1P concept is set to p_1, as computed from $r_p=p_2/p_1$. In case of the HPES-3P system, the intermediate pressure p_M for a given value of p_L is determined using Eqt 7. As noted in Fig. 4a, higher values for p_L lead to more compact HPES-2P and HPES-3P designs, resulting in higher volumetric densities. However, higher p_L values lead

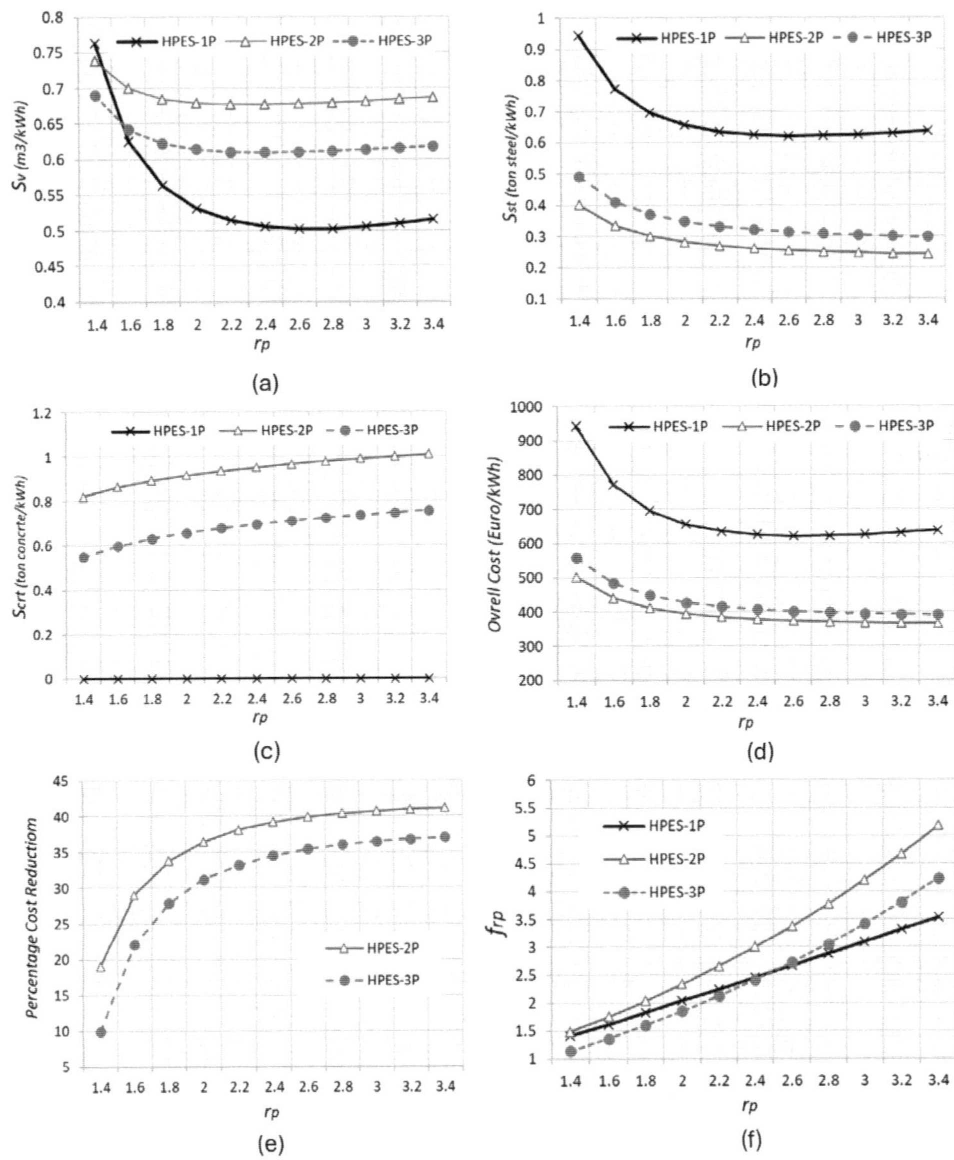

Figure 2. Influence of the pressure ratio r_p for three HPES concepts

to increased steel requirements for the pressure vessels, thus increasing the weight of the design (Fig. 3b). This in turn reduces the mass requirements for concrete, as shown in Fig 3c. It is observed that the HPES-3P system has smaller volumetric requirements than the HPES-2P system for all p_L values considered.

The percentage cost reductions achieved in the pressure containment through the HPES-2P and HPES-3P concepts, with respect to the HPES-1P concept, are shown in Fig. 4d. It is shown that the highest cost reduction (around 35 – 40%) is achieved at p_L around 20 – 25 Bar. Beyond such pressures, no further appreciable cost reductions are expected. Figure 4e shows the variation of

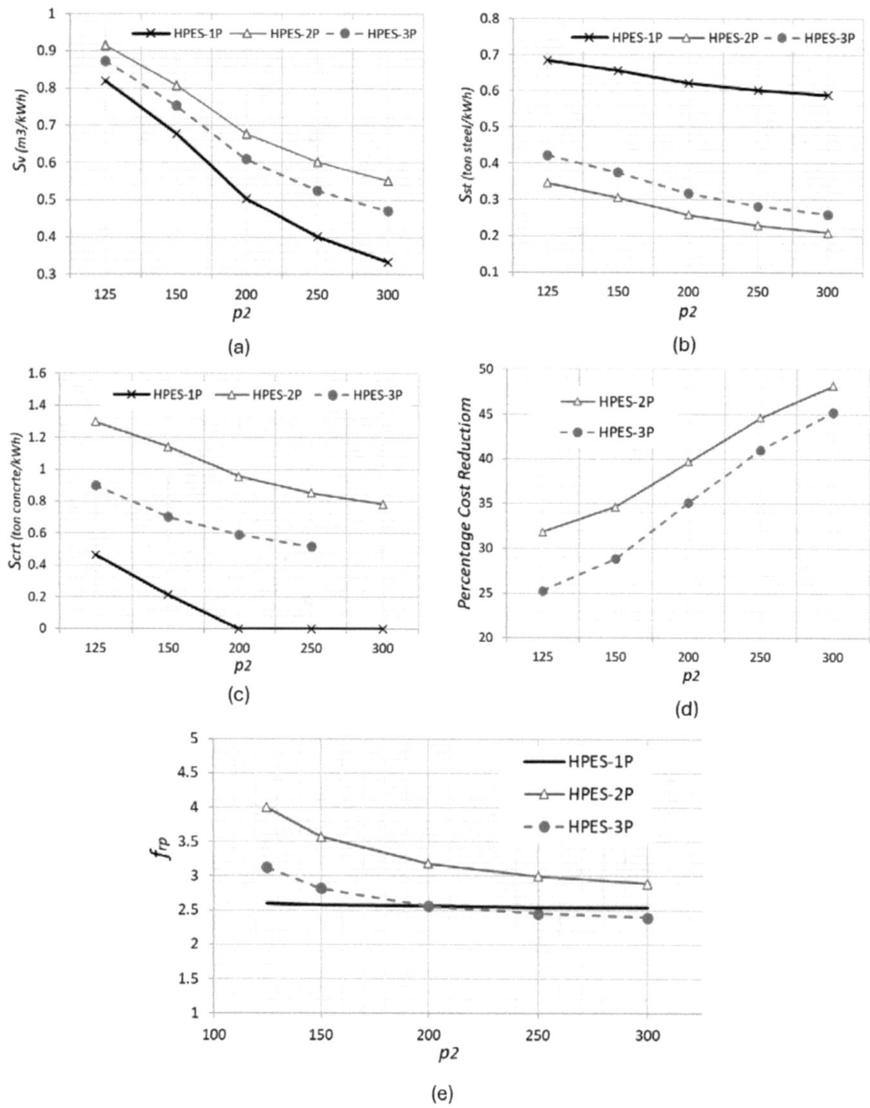

Figure 3. Influence of the HPV operating pressure for three HPES concepts

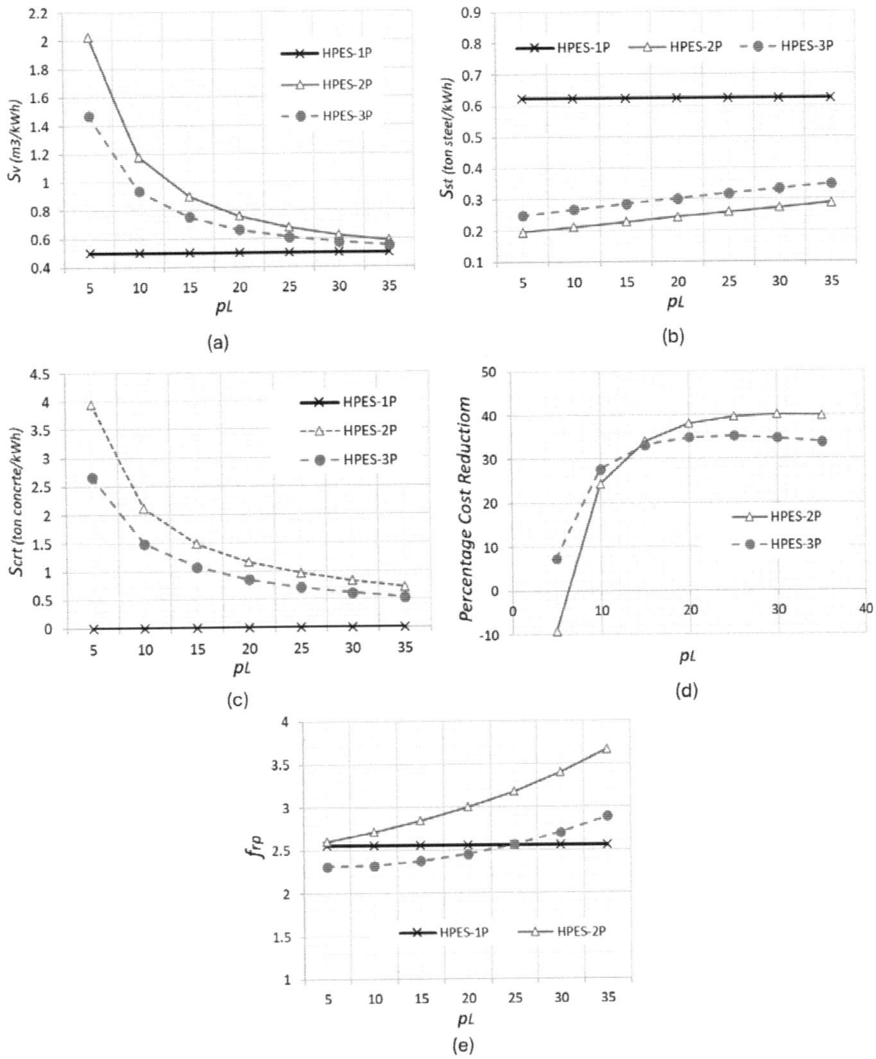

Figure 4. Influence of the pre-charge pressure of the LPV in the HPES-2P and -3P concepts.

pressure difference factor, f_{rp}. Increasing p_L leads to higher factors, yet that of the HPES-2P increases more rapidly. This is expected to have a negative impact on the AEM due to the wider operating pressure range that the machine will be exposed to. On the other hand, the HPES-3P maintains a low value, that is only marginally different than that for the HPES-1P system. This highlights the benefit of using a second liquid piston accumulator as shown in Fig. 1c.

6. Conclusions

This study compared the steel and concrete requirements for three different HPES concepts for subsea deployment. The following conclusions could be drawn:

- The split-type HPES-2P and HPES-3P concepts can reduce significantly the steel requirements for the pressure containment (in terms of kg of steel per kWh) when compared to the simple HPES-1P concept. However, the two concepts will require concrete anchors to overcome buoyancy and have them installed on the seabed. Such anchoring requirements would theoretically not be necessary for the HPES-1P concept for a design peak working pressure of

200 Bar or higher. On the other hand, the other two concepts will require concrete even at design pressures exceeding 300 Bar.

- Despite having the highest steel requirements per unit of stored energy, the HPES-1P concept remains the most compact solution for subsea deployment due to the highest volumetric density.
- The potential cost reductions in the pressure containment for the HPES-2P and HPES-3P concepts, as compared to the HPES-1P concept, are estimated to be around 40% and 35%, respectively. This accounts for both the steel and concrete requirements. Such reductions are achievable for a pre-charge pressure for the LPV of 20 – 25 Bar and an HPV pressure ratio >2.6.
- The pressure containment of HPES-2P concept is the most cost viable solution out of the three HPES options. However it leads to an increased operating pressure window for the AEM, thus impacting the energy conversion efficiency. The HPES-3P concept can provide a smaller operating pressure window for the AEM, while still achieving a 35% cost reduction for the pressure containment with respect to the HPES-1P concept.

Further work should explore the HPES-3P concept in more detail, with more comprehensive modelling to investigate the performance of the energy conversion turbomachines required and the thermodynamic efficiency achieved when operating in a subsea environment. More detailed modelling is also required to account for other costs, as for example transport and installation costs for each HPES concept.

Acknowledgements

The work forms part of the DuaL-HPES Project being supported through the Maritime Seed Award 2023, with funding from Transport Malta.

References

[1] Zhiwen Wang, Rupp Carriveau, David S.-K. Ting, Wei Xiong, Zuwen Wang. A review of marine renewable energy storage. *International Journal of Energy Research*. 2019; 43:6108–6150, doi-org.ejournals.um.edu.mt/10.1002/er.4444.

[2] Arellano-Prieto Y, Chavez-Panduro E, Salvo Rossi P, Finotti F. Energy Storage Solutions for Offshore Applications. *Energies*. 2022; 15(17):6153. https://doi.org/10.3390/en15176153

[3] Bennett JA, Simpson J.G., Qin C, Fittro R, Koenig GM , Clarens AF, and Loth E. Techno-economic analysis of offshore isothermal compressed air energy storage in saline aquifers co-located with wind power. *Applied Energy*. 2021; 303, 117587, https://doi.org/10.1016/j.apenergy.2021.117587.

[4] Hurley W, Orthmann A, Lieberman P and Enis B. Engineering and Cost Study of an Offshore Wind Farm Compressed Air Energy Storage System. Offshore Energy and Storage Summit (OSES), BREST, France, 2019. pp. 1-11, doi: 10.1109/OSES.2019.8867346.

[5] Dick C, Puchta M, Bard J, StEnSea – Results from the pilot test at Lake Constance. *Journal of Energy Storage*.2021; 42, 103083, https://doi.org/10.1016/j.est.2021.103083

[6] Nienhuis RM, van Rooij M, Prins WA, Jayawardhana B, Vakia, AI. Investigating the efficiency of a novel offshore pumped hydro energy storage system: Experimental study on a scale prototype. *Journal of Energy Storage*. 2023; 74, Part A.

[7] Pimm AJ, Garvey SD, de Jong M. Design and testing of Energy Bags for underwater compressed air energy storage. *Energy*. 2014; 66: 496-508, https://doi.org/10.1016/j.energy.2013.12.010.

[8] Swinfen-Styles L, Garvey SD, Giddings D, Cárdenas B, Rouse JP. Analysis of a Wind-Driven Air Compression System Utilising Underwater Compressed Air Energy Storage. *Energies*. 2022; 15(6):2142. https://doi.org/10.3390/en15062142.

[9] Hunt JD, Zakeri B, Giulietti de Barros A, Filho WL , Marques AD, Barbosa PSF, Smith Schneider P, Farenzena M, Buoyancy Energy Storage Technology: An energy storage solution for islands, coastal regions, offshore wind power and hydrogen compression. *Journal of Energy Storage*. 2021; 40:102746.

[10] Silva J, Maia Alves J, Sensitivity analysis of levelized cost of hydro-pneumatic electricity storage for grid support applications. *Renewable Energy*. 2023; 219, Part 1: 119285, https://doi.org/10.1016/j.renene.2023.119285

[11] Saadat M, Shirazi FA, Li, PY. Modelling and control of an open accumulator CAES system for wind turbines. *Applied Energy*. 2015;137: 603-616. https://doi.org/10.1016/j.apenergy.2014.09.085.

[12] Abu-Heib, A. Ahmad, Kassaee, S, Chen, Y, Smith, B. Pumped Storage Hydropower Augmented with Pressurized Air: The Ground-Level Integrated Diverse Energy Storage (GLIDES) System — GLIDES System Configurations and Use Cases. United States: N. p., 2022. United States. https://doi.org/10.2172/1891390.

[13] Zhao, P. Lai Y, Zhang S, Xu W, Wang J, Dai Y. Near constant discharge performance analysis of a dual accumulator configuration quasi-isothermal compressed gas energy storage based on condensable gas. *Journal of Energy Storage*. 2020, 32: 101945. https://doi.org/10.1016/j.est.2020.101945.

[14] Buhagiar D, Sant T, Farrugia RN, Aquilina L, Farrugia D, Strati FN. Small-scale Experimental Testing of a Novel Marine Floating Platform with Integrated Hydro-pneumatic Energy Storage. Journal of Energy Storage. 2019; 24: 100774. https://doi.org/10.1016/j.est.2019.100774.

[15] Barbour E, Pottie DL, Adiabatic compressed air energy storage technology, *Joule*, 2021; 5, Issue 8: 1914-1920.

[16] Gouda EM, Fan Y, Benaouicha M, Neu T, Luo L. Review on Liquid Piston technology for compressed air energy storage. Journal of Energy Storage, 2021; 43: 103111. https://doi.org/10.1016/j.est.2021.103111.

[17] Cutajar C, Sant T, Farrugia R N., Buhagiar D, A Software Tool for the Design and Operational Analysis of Pressure Vessels used in Offshore Hydro-pneumatic Energy Storage, Journal of Energy Storage, 2021; 40: 102750, ISSN 2352-152X, https://doi.org/10.1016/j.est.2021.102750.

The Effect of Wind Turbines' Load Factor on the Overall Cost of Electricity

Bruno Cardenas[1*], Seamus Garvey[1], Zahra Baniamerian[1] and Ramin Mehdipour[1]

[1] Department of Mechanical, Materials and Manufacturing Engineering, University of Nottingham, United Kingdom

*E-mail: Bruno.Cardenas@nottingham.ac.uk

Abstract. The load factor of wind turbines has a strong effect on the wider electricity system. As the rated wind speed of turbines reduces, the load factor increases. This allows wind farms to generate constant power for longer periods of time, which in turn improves the match between the profiles of electricity demand and renewable generation and reduces the need for energy storage. The behaviour of the total cost of electricity is driven largely by the levelized cost of wind power, which has a minimum at a rated speed of 12 m/s. Faster rated speeds (>12 m/s) lead to higher electricity costs due to the lower load factors of the turbines. At slower rated speeds (<12 m/s) the higher capital cost of the turbines overshadows the improved load factors, leading to increased electricity costs.

1. Background and context

Grid-scale energy storage will be needed to enable future zero-carbon electricity systems integrate large shares of renewable generation [1-4]. Energy storage helps to balance the grid by storing 'excess' renewable electricity produced during times of low demand and discharging it back into the grid during times of high demand and/or low generation [5-7].

Considerable research has been devoted to quantifying the amount of energy storage capacity that future electricity systems will require [7-12]. In previous work, the authors have carried out an assessment of the storage capacity that the UK will need to decarbonise its electricity system [13]. The authors have also explored how to combine different storage technologies (hydrogen caverns, compressed air energy storage (CAES) and Li-ion batteries) to provide the capacity required at the lowest possible cost for the system [14]. The two studies mentioned above have used UK electricity demand and renewable generation data spanning several years. The studies have analysed the system-level effect of different variables, such as: the penetration level of renewables, the mix between wind and solar PV, the amount of over-generation (i.e. curtailment) in the system and the distribution of the storage duty between different technologies with the aim of determining the system configuration that minimises the overall cost of electricity (£/MWh).

Results indicate that the UK will need an energy storage capacity of approximately 66 TWh to fully decarbonise its electricity network. The optimum mix of renewables for the country is 85% wind and 15% solar PV. There is also 15% of over-generation (and curtailment), which helps improving the real-time consumption of energy and minimising the need for storage [14]. In this scenario, underground hydrogen caverns provide ~ 55 TWh of long duration storage capacity (approx. 140 days) while CAES systems provide 11 TWh of medium duration capacity (approx. 10

days). Li-ion batteries provide a much smaller 0.2 TWh of storage. However, this capacity is essential to deal with short but very frequency imbalances in the grid.

The potential impact of the load factor of wind turbines on the requirement for energy storage has not been explored yet. The load factor of a wind turbine is defined as the ratio between the energy produced over a period of time and the energy that could have been generated over the same period if the turbine had worked the whole time at its maximum capacity. In the previous studies, the turbines' load factor has been implicit in the power generation data. However, the data also includes factors that are external to the turbines, such as curtailment due to grid congestion or low demand and transmission losses. This paper explores whether improving the load factor of wind turbines could bring additional benefits from a system point of view.

2. Wind speed data and power generation profiles

As mentioned, previous studies have used historical data on electricity demand and renewable generation. Demand data has been sourced from [15,16], while solar PV generation data was obtained from Renewables Ninja [17]. This study uses the same demand and solar PV generation data as previous work. However, wind power generation data cannot be used as it does not allow to see the effect of changing the turbines' load factor.

Instead, we use wind speed data from the 15 largest offshore wind projects in the UK. These include existing projects but also projects that are under-development or that have an agreement in place [18]. Using Renewables Ninja [17], we obtained 12 years of wind speed data for each of the 15 different locations.

To convert this data into a power generation profile we need the power curve of a turbine. The Vestas V164, which has a rated capacity of 9.5 MW, was selected as the reference turbine for this work. This turbine is the top 1 model in the UK by total installed capacity and the top 3 model by number of installed turbines. This reference turbine has a cut-in speed of 3.5 m/s, a rated wind speed of 14 m/s and a cut-off speed of 25 m/s. Below the rated speed, the output of the turbine is only a fraction of its nominal capacity. Above the rated speed (and below the cut-off) the turbine produces its nominal output [19]. Based on this power curve, we created several other power curves for different rated wind speeds, as Figure 1 shows.

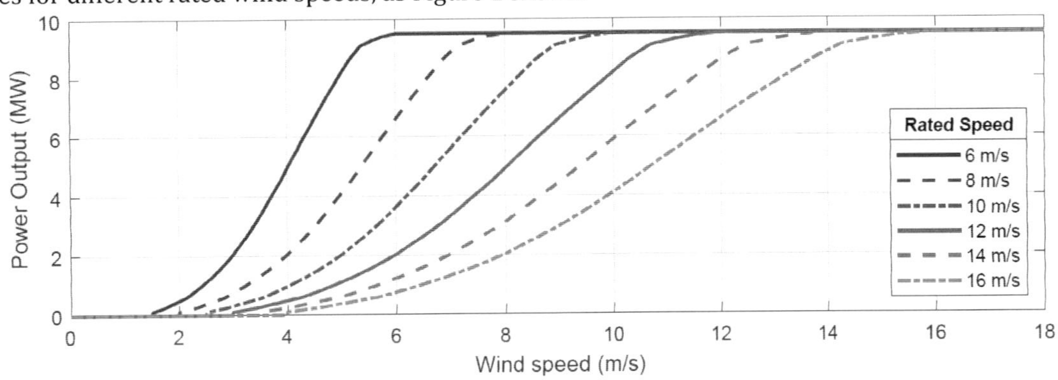

Figure 1. Power curves for turbines with different rated wind speeds. Reference curve is based on a rated speed of 14 m/s. In all cases the cut-off speed is 25 m/s.

We assume that for all the different power curves, the ratio between the real output of the turbine at the rated speed and its theoretical maximum stays constant. In all cases, the real output is 9.5 MW; therefore, the theoretical maximum also stays constant for all the different rated speeds. We also assume that for all the different power curves the ratio between the rated speed and the cut-in speed remains constant. We can calculate the swept area (A) of the turbine by means of Eq. (1), where ρ is the density of the air, v is the wind speed and φ is a coefficient of power. In this study we use a value for φ of 0.59, which is the Betz limit. We find that for a speed of 14 m/s the swept area is 21,125 m^2 (actual dimension of the Vestas V164). This increases by a factor of 8 when the rated wind speed reduces by a factor of 2.

$$P_{max} = \frac{1}{2} \cdot A \cdot \rho \cdot v^3 \cdot \varphi \tag{1}$$

We can create wind power generation profiles, like the ones shown in Figure 2, by combining the wind speed data of the 15 offshore locations with the power curves. The power generation profiles are normalised to facilitate their use. Figure 2 shows in a very clear way the effect that the rated speed of the turbines has on their load factor. As the rated wind speed reduces, the turbines produce a higher output for a greater fraction of the time. Figure 3 shows how the load factor and the blade length of the turbines increases as the rated wind speed reduces.

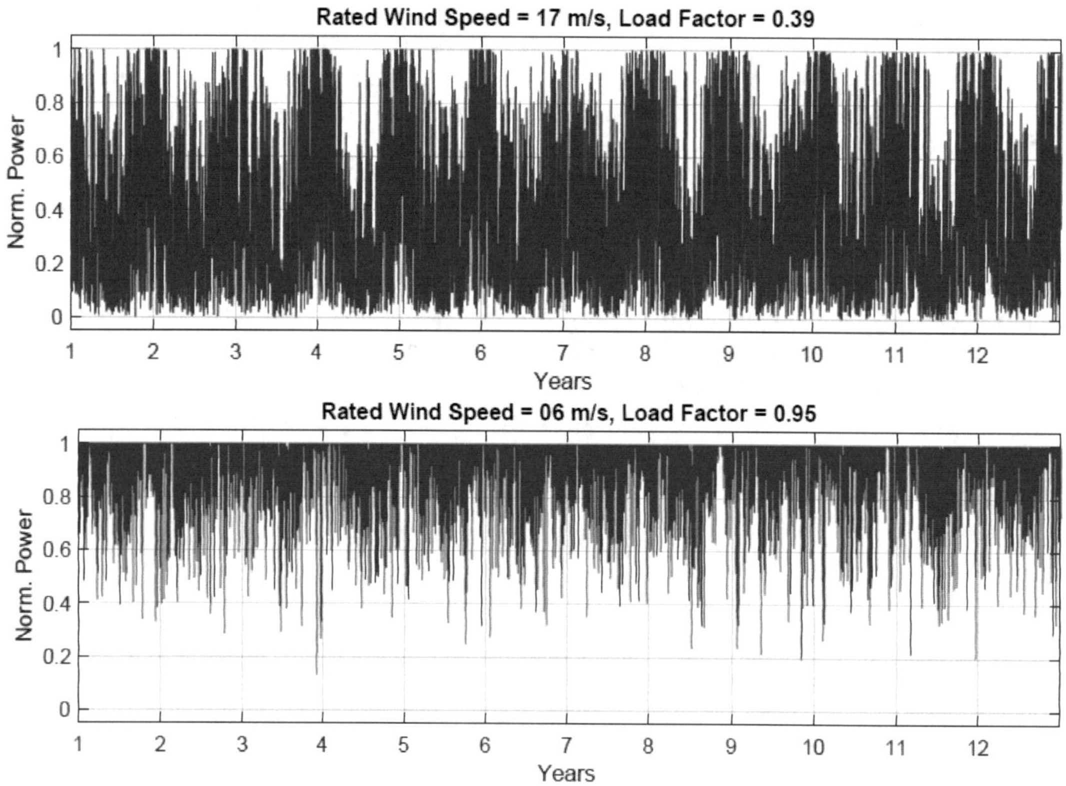

Figure 2. Wind power generation profiles for different rated wind speeds: 17 m/s (top) and 6 m/s (bottom)

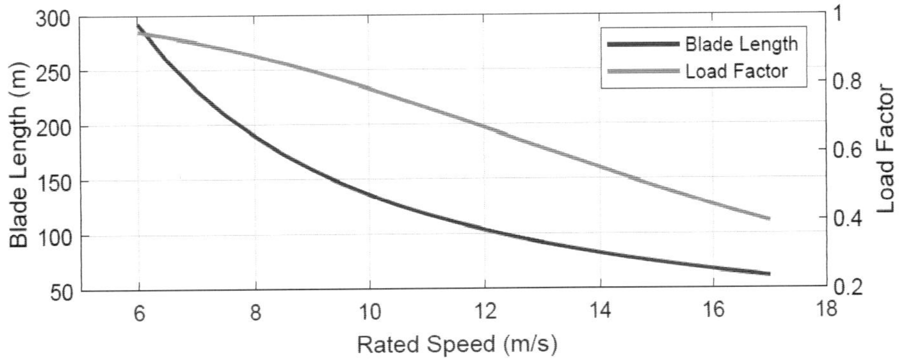

Figure 3. Load factor and blade length of a wind turbine with respect to its rated wind speed

2.1 Cost of wind power as a function of rated speed

The performance of a wind turbine improves as its rated speed reduces, due to a higher load factor. However its blades become longer, which translates into a higher capital cost. In this section we discuss how the cost of wind power for the different turbines considered (i.e. different rated speeds) is calculated.

A.B. Abrahamsen *et al* [20]. provided an equation that relates the mass (m) of a wind turbine's blades (in tons) to their length (L).

$$m = 129 \cdot 10^{-5} \cdot L^{2.32} \tag{2}$$

We assume that the cost per MWh of electricity produced by the reference turbine (rated speed 14 m/s) is 40 £/MWh. This aligns with current costs of offshore wind in the UK [21]. From Figure 3 we know that the load factor of this turbine is 0.55; therefore, we can calculate how much energy it will produce over the 12-year period. The capital cost of this turbine can be calculated simply by multiplying the total energy produced by the cost per MWh.

According to a study published by the Offshore Renewable Energy Catapult [22], the capital cost of wind turbines accounts for 29.5% of the cost of a wind farm (i.e. project cost). This includes the nacelle (11.7%), rotor (5.6%), tower (1.9%) and other turbine costs (10.3%). The remaining 70.5% comprises costs such as the foundation of the turbine, offshore substation, cables and installation, maintenance and service, operations and decommissioning. Based on this we know that 18.9% of the cost of a wind turbine is attributable to its rotor. This breakdown allows us to calculate the cost of the rotor (for the reference turbine) and the cost per ton of blade material (£/ton). Having determined the cost per ton of blade material, we can calculate the rotor cost for all the other cases (i.e. different rated speeds).

The cost of the different turbines for the different rated speeds analysed can be calculated by dividing the cost of the rotor by a ratio Z, which represents the fraction of the turbine's cost that is attributable to the rotor. As mentioned, in the reference case (rated speed = 14m/s) the ratio Z was 0.189. We assume that the ratio Z goes from 0.12 for a rated speed of 17 m/s up to 0.38 for a rated speed of 6 m/s. As the rated wind speed reduces and the turbine becomes bigger, the rotor accounts for a greater fraction of the cost. Larger rotors induce greater loads on the support structure (i.e. tower). This is not factored in the cost breakdown and it is therefore a limitation of this study.

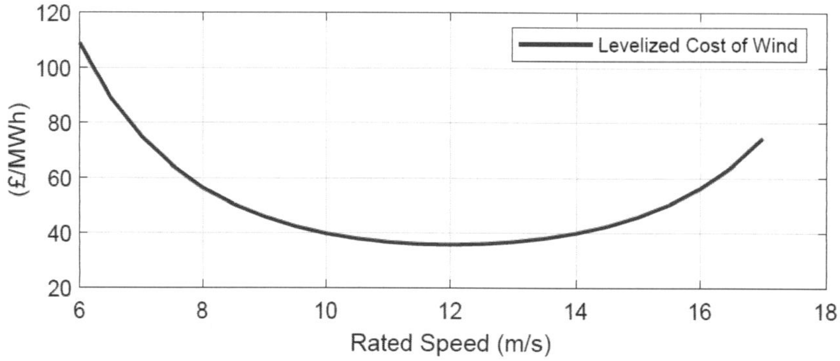

Figure 4. Effect of the rated speed of a wind turbine on the levelized cost of the power produced

The overall project cost for the different rated speeds can be calculated by dividing the cost of the turbine by a ratio *Q*, which represents the fraction of the overall project cost (*CapEx* and *OpEx*) that is attributable to the wind turbine. In the reference case this ratio was 0.295. We assume that the ratio *Q* goes from 0.18 for a rated speed of 17 m/s up to 0.6 for a rated speed of 6 m/s. As the rated wind speed reduces and the turbine becomes more expensive (due to a larger rotor), it accounts for a greater proportion of the overall project cost.

The levelized cost of wind power can be calculated by dividing the overall project cost for each of the different rated speeds by the total energy produced during the period analysed. Figure 4 shows how the levelized cost of wind power varies with respect to the rated wind speed of the turbines. The levelized cost of wind power is high when rated speeds are high due to the low load factors achieved. The levelized costs of wind power seen at slow rated speeds are also high. This is owed to the fact that the capital cost of the turbines overshadows the improvement in performance. The levelized cost of wind is minimised at a rated speed of 12 m/s.

3. Methodology: Modelling the requirement for energy storage

Different rated wind speeds (***V***) are evaluated. For each one of those, we carry out a combinatorial experiment with different system variables to find the "configuration" that minimises the overall cost of electricity. These variables are the mix of renewables (***R***), the amount of over-generation in the system (***Ω***) and the mix of energy stores (***X***).

The procedure followed to analyse the different cases (combination of ***V***, ***R***, ***Ω*** and ***X***) is very similar to that used in some of the author's previous work. A comprehensive explanation of the modelling tool and the procedure followed can be found in [13,14].

The modelling tool use is a time-stepping algorithm developed in MATLAB. This algorithm receives three (normalised) time-series expressed in power units: 1) an electricity demand profile, 2) a profile of solar PV power generation and 3) a profile of wind power generation (as shown in Figure 2). These time series have an hourly resolution and span a period of 12 years (i.e. 105,120 data points). The model also receives three user inputs, which are the system variables. The first one is a ratio ***R***, which represents the contribution of wind power to the total energy generated. The contribution of solar power is simply 1-***R***. The second variable ***Ω***, represents the amount of over-generation (and energy curtailment) that is allowed in the system. This is expressed as a fraction of the total energy demand. The third variable is a ratio ***X***, which indicates the ratio between how much energy is sent to CAES with respect to the total amount of energy that needs to be sent/passed through storage. The hydrogen caverns receive 1-***X***.

For a given combination of parameters, the model multiplies the normalised profile of electricity demand by a factor *f* so that the total demand over the 12 years is 4020 TWh. This equates to an average annual demand of 335 TWh, which is representative of current consumption levels. The normalised profiles of wind and solar PV generation are also amplified so that each resource produces the correct amount of electricity (based on $\boldsymbol{R}$ and $\boldsymbol{\Omega}$). Renewables will generate enough energy to meet demand, plus an additional amount given by $\boldsymbol{\Omega}$ (which is curtailed) and a further extra to compensate for storage losses.

After knowing how much renewable energy is needed in the system, the algorithm proceeds to creating a profile of *Net Demand* (demand minus generation) which is effectively a duty cycle for the energy stores. The algorithm splits the profile of *Net Demand* into two separate profiles, a work cycle for the long-duration energy stores (underground H_2 caverns) and a work cycle for the medium duration store (CAES). The algorithm tries to minimise the storage capacity provided by CAES as it is more expensive per kWh than the capacity provided by H_2. In parallel, the algorithm minimises the rated powers (charge and discharge) of the hydrogen store as this is more expensive per kW than CAES. The two work profiles replicate the profile of *Net Demand* perfectly and their sign is equal to the sign of *Net Demand* at all points in time, which avoids counterflow of energy and the stacking up of inefficiencies by discharging one store to charge the other.

Figure 5 shows an example of the work profiles created by the tool. Based on the work profiles we can calculate the capacity of the two energy stores as well as their rated powers. In turn, this enables us to calculate the overall system cost or 'total cost of electricity' (*TCoE*).

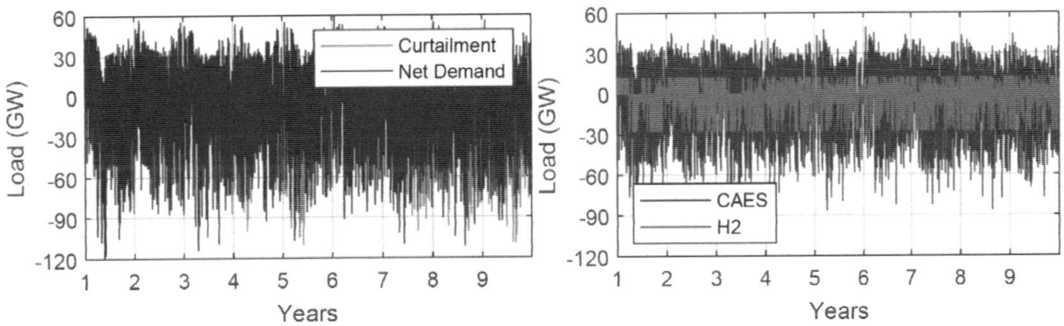

Figure 5. Examples of work profiles produced by the modelling tool. Left: Curtailed profile of net demand. Right: Work profiles for the H_2 and CAES stores

4. Results and discussion

Figure 6 shows how the profile of *Net Demand* changes as the rated speed of wind turbines varies. The load factor of wind turbines improves as their rated speed reduces, which means that demand can be fully met in real time more of the time. This reduces the need for energy storage. Each of the *Net Demand* profiles shown in the figure considers the optimum values of $\boldsymbol{R}$, $\boldsymbol{\Omega}$ and $\boldsymbol{X}$ for the specific rated wind speed. The optimum mix of wind and solar ($\boldsymbol{R}$) for the different rated speeds was determined separately prior to the multi-variable exploration. As the rated speed of turbines reduces, the system incorporates more wind into the mix. At a rated speed of 17m/s, the share of wind power is 78% (the other 22% is supplied by solar PV). This increases to 88% when the rated speed reduces to 14 m/s. At rated wind speeds of 11 m/s and below, wind power supplies 100% of the energy.

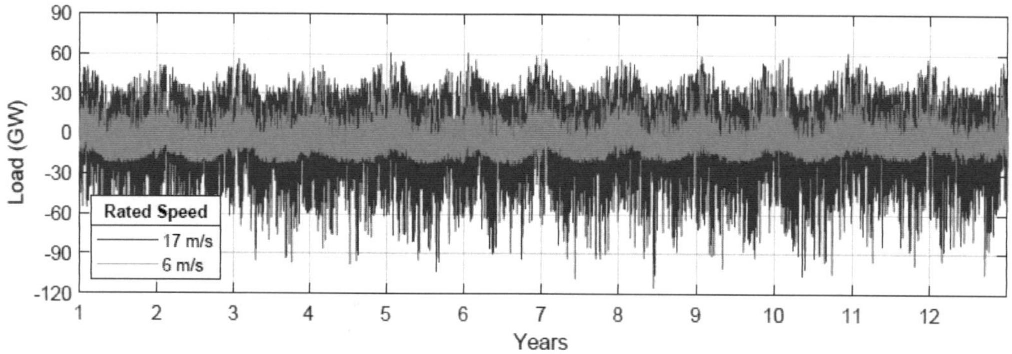

Figure 6. Effect of the rated wind speed of turbines on the shape of the profile of *Net Demand*

Figure 7 shows how the total cost of electricity (*TCoE*) varies with respect to the turbine's rated wind speed. We can see that the *TCoE* is high at both ends of the range of rated speeds. Fast rated speeds lead to an increased *TCoE* because the cost of wind power is high in this region, due to the low load factors of the turbines (See Fig. 3). On the other hand, slow rated speeds also lead to a high *TCoE* because here, the cost of wind power is high too. As explained before, the size and cost of wind turbines increases as their rated speed reduces. The increase in capital cost overshadows the improvement in load factor, leading to higher electricity costs. The *TCoE* is minimised at a rated speed of 11m/s (50.3 £/MWh). It is important to note that each point shown in Figure 7 considers the optimum combination of $\boldsymbol{R}$, $\boldsymbol{\Omega}$ and $\boldsymbol{X}$ for the specific wind speed.

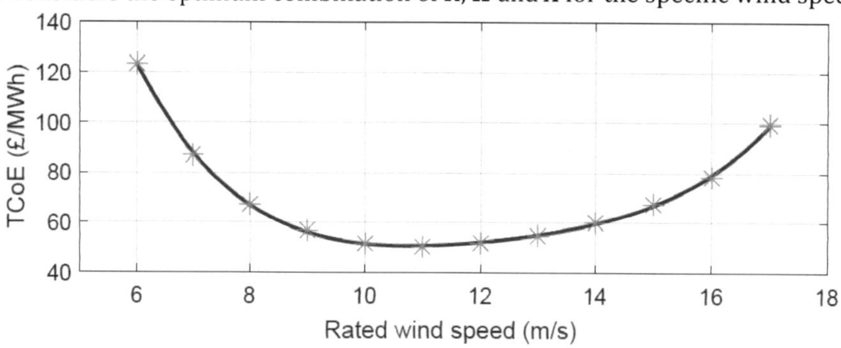

Figure 7. Variation in the total cost of electricity with respect to the rated speed of wind turbines

The total cost of electricity (*TCoE*) is calculated through Eq. (3). The *TCoE* encompasses the cost of electricity generation and of the energy storage provisions.

$$TCoE = \left(C_W + C_S + C_{CAES} \cdot (\tau/\lambda) + C_{H2} \cdot (\tau/\lambda)\right) \cdot E_D^{-1} \tag{3}$$

In Eq. (3), C_W and C_S are the total costs of the electricity generated from wind and solar PV, respectively. We assume that solar PV electricity costs 60 £/MWh [23]. C_{CAES} and C_{H2} are the capital costs of the energy stores, τ is the length of the period analysed (in years) and λ is the useful life of the assets. E_D is the total electricity demand over the 12 years (4020 TWh). The capital cost of the energy stores is calculated by means of Eq.(4), where α is the cost per kWh of storage capacity, β is the cost per kW of charging power and γ is the cost per unit of rated discharge power. The economic figures used are given in Table 1.

$$C_{store} = \alpha \cdot S + \beta \cdot abs(P_c) + \gamma \cdot P_d \tag{4}$$

Figures 8 to 11 explain how different factors that influence the total cost of electricity vary with respect to the rated speed (and load factor) of wind turbines. Figure 8 shows the energy storage capacities provided by CAES (left) and underground hydrogen caverns (right). It can be observed that the hydrogen storage capacity increases as the ratio ***X*** decreases because a greater fraction of the energy that needs storage is being sent to the H_2 store. Conversely, the storage capacity provided by CAES increases with ***X***, because CAES is servicing a greater fraction of the storage duty. The overall capacity in the system is the sum of two stores' capacities. The capacities of both stores reduce as the rated wind speed decreases. The reason for this is that as the rated speed of wind turbines reduces and their load factor improves, more renewable energy is consumed in real time and less energy needs to pass through storage.

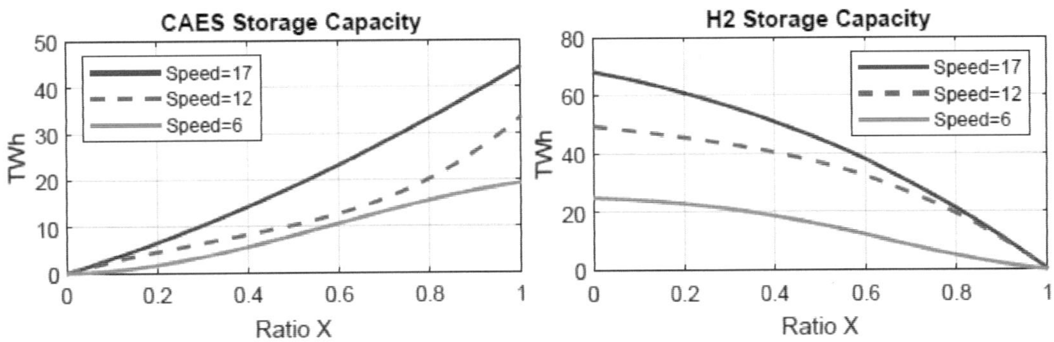

Figure 8. Effect of the ratio ***X*** (distribution of storage duty) and the rated speed of wind turbines on the storage capacities provided by CAES (left) and underground hydrogen caverns (right)

Figure 9 shows how the charging and discharging powers of the hydrogen store change with respect to the rated wind speed. The charging powers are expressed as negative values for consistency, given they are dictated by the negative part of the *Net Demand* profile. We can see that the magnitudes of both powers, charging and discharging, increase as the ratio ***X*** moves towards zero because the hydrogen store handles a greater share of the storage duty. We can also see that both rated powers reduce in magnitude as the rated wind speed reduces. This is because slower wind speeds lead to better load factors, which reduces the peak-to-peak distance of the *Net Demand* profile (see Fig. 6).

Table 1. Economic figures used for the calculation of the "Total Cost of Electricity"

Parameter	CAES	Hydrogen
Storage capacity cost (α)	3.5 £/kWh [24]	0.67 £/kWh [25]
Charging power cost (β)	300 £/kW [24]	1100 £/kW [26]
Discharging power cost (γ)	300 £/kW [24]	450 £/kW [27]
Roundtrip efficiency (η)	70% [28]	45 % [29,30]
Service life (λ)	30 Years [31]	30 Years [31,4]

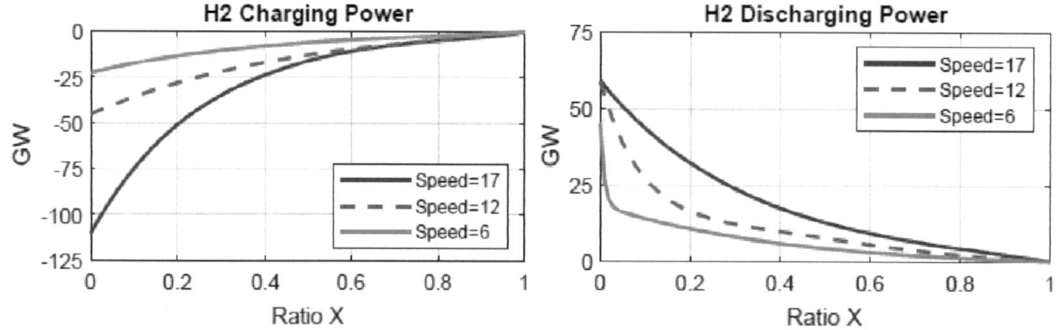

Figure 9. Behaviour of the rated charge and discharge powers of the hydrogen store (left and right, respectively) with respect to the distribution of the storage duty (***X***) and the rated wind speed

The behaviour of the charging and discharging powers of CAES with respect to changes in the ratio ***X*** and the rated wind speed is shown in Figure 10. We see a very similar behaviour to that of the hydrogen stores. The magnitudes of both rated powers (charge and discharge) increase as ***X*** tends to 1 because more energy is passing through CAES. We can also see in Figure 10 that the magnitudes of both, the rated charge and discharge powers of CAES, decrease as the rated speed of wind turbines reduces and their load factor improves.

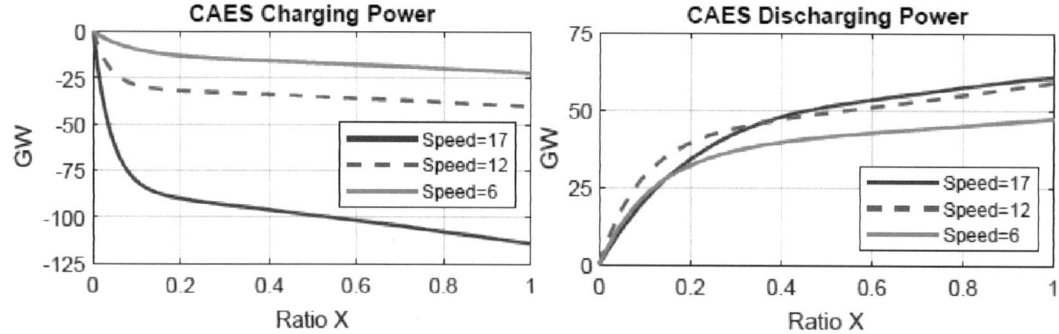

Figure 10. Behaviour of the rated charge and discharge powers of CAES (left and right, respectively) with respect to the distribution of the storage duty (***X***) and the rated wind speed

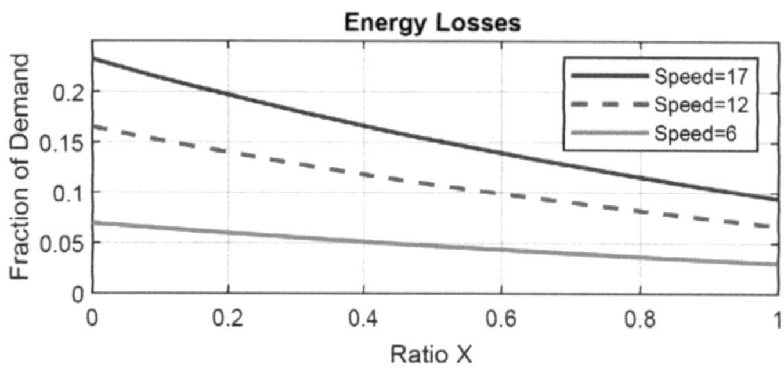

Figure 11. Storage losses with respect to distribution of the storage duty (X) and the rated wind speed

Figure 11 shows how the energy losses in the system vary depending on the distribution of the storage duty between CAES and hydrogen (***X***) and the rated speed of wind turbines. Here, losses are shown as a fraction of total demand (4020 TWh). Losses increase as ***X*** tends to zero because more energy is passed through the hydrogen stores, which have a lower roundtrip efficiency. Energy losses decrease as the rated wind speed reduces because, as mentioned, more energy is consumed in real time and less energy passes through storage.

Figures 8 to 11 have shown that most of the cost-driving parameters (storage capacity, rated charge/discharge power and energy losses) reduce monotonically as the rated speed of wind turbines reduces. The behaviour of the *TCoE* (with a minimum at v=11 m/s) is determined mainly by the levelized cost of wind power, which is a trade-off between the capital cost of a turbine and its load factor.

The reference turbine used for this study has a rated speed of 14 m/s and achieves a load factor of 0.552. Most turbines in the UK have rated speeds in this range. However, if we look at a timeseries of similar length obtained from the UK's electricity balancing system [15,16] we find that wind power has a load factor of ~0.36 (after accounting for the increase in installed capacity over time). The reason for this difference could be that the profile of wind power generation from the grid operator includes downtime due to maintenance, curtailment due to grid congestion issues, transmission losses, etc. whereas a profile of power generation based on wind speed can be seen as an ideal scenario or an 'upper limit'.

The difference in the load factors of the profiles leads to different estimations of the energy storage requirements and ultimately to considerably different figures for the *TCoE*. This very important consideration has been brought to light by the present study. We will explore this in depth in future work with aims to provide guidance to other researchers conducting energy system modelling. We believe that using wind power generation profiles based on historical data source from the grid operator may led to overestimations in the energy storage capacity needed. For example, in this study we find that when the rated speed is 14 m/s and the load factor is 0.552, the *TCoE* is approx. 60 £/MWh. In our previous work, which uses wind power generation data (instead of speed data) we have found a *TCoE* in the region of 71 £/MWh (excluding battery costs).

Due to this discrepancy, the results reported in this paper in terms of cost of electricity (*TCoE*) should be treated as indicative "lower limits" rather than exact figures. Nevertheless, the general trends presented in the paper are not affected by the discrepancy between wind power profiles. The study has provided insight into how the rated speed and load factors of wind turbines influence the overall electricity system and the cost per MWh, which was our main research question.

5. Concluding remarks

This paper explores the effect that the load factor of wind turbines can have on the electricity system of the UK, particularly on the need for energy storage and the total cost per MWh of electricity.

A time-stepping algorithm developed in MATLAB is used to model the electricity grid. This single node model aims to maintain an energy balance in the grid. The algorithm sizes two centralised energy stores to minimise energy curtailment and to ensure that demand is always met. Underground hydrogen caverns and CAES are used for the long and medium duration energy storage duties, respectively. The model uses three timeseries: a profile of electricity demand, a profile of solar PV power generation and a profile of wind power generation. In this study we use historical wind speed data from the 15 largest offshore sites in the UK to create profiles of wind power generation based on different turbine power curves with different rated speeds.

The load factor of a wind turbine depends on its rated speed. The study has shown that the load factor of wind turbines (given by their rated speed) has a noticeable effect on the overall electricity system. Current offshore wind turbines have rated speeds in the order of 14 m/s.

Results indicate that the total cost of electricity (*TCoE*) can be minimised with a rated speed of 11 m/s which translates to a load factor of 0.728.

At rated wind speeds slower than 11 m/s, the improvement in the load factor of turbines is overshadowed by their capital cost. This leads to an increased *TCoE*, despite reductions in the other cost-drivers of the system such as the energy storage capacity, the rated power of the stores and energy losses due to storage. Rated wind speeds higher than 11 m/s also lead to an increased *TCoE*. As the rated speed of a wind turbine increases, both its capital cost and load factor reduce. However, the reduction in the load factor exceeds the reduction in capital cost, leading to a higher cost per MWh of wind-based electricity. In addition to this, a reduced load factor causes an increase in all the other cost-driving factors: the storage capacity required, the rated charging/discharging powers of the energy stores and the energy losses due to storage. All of this combined leads to an increased *TCoE*.

References

[1] Schmidt O and Staffell I. 2023 *Monetizing energy storage: A toolkit to assess future cost and value*. Oxford, United Kingdom: Oxford University Press.
[2] Child M, Bogdanov D, Breyer C. 2018 *Energy Procedia* **155** 44-60
[3] Schill W P. 2020 Joule **4** 2059–2064
[4] Dowling J A, Rinaldi K Z , Ruggles T H, *et al.* 2020 *Joule* **4** 1907-1928
[5] Alexopoulos D K, Anastasiadis A G, Vokas G A, Kaminaris S D, 2021 *Energy Reports* **7** 33-50
[6] Beaudin M, Zareipour H, Schellenberglabe A, Rosehart W. 2010 *Energy Sustain. Dev.* **14** 302-314
[7] Cebulla F, Naegler T, Pohl M. 2017 *J. Energy Storage* **14** 211–223
[8] Grunewald P, Cockerill T, Contestabile M, Pearson P. 2011 *Energy Pol.* **39** 4807-4815
[9] Heide D, Greiner M, von Bremen L, Hoffmann C. 2011 *Renew. Energy* **36** 2515-2523
[10] Denholm P, Hand M. 2011 *Energy Pol* **39** 1817-1830
[11] Solomon A A, Kammen D M, Callaway D. 2014 *Appl. Energy* **134** 75-89
[12] Solomon A, Child M. 2017 *Energy Procedia* **135** 283-293
[13] Cardenas B, Swinfen-Styles L, Rouse J, Hoskin A, Xu W, Garvey S D. 2021 *Renew. Energy* **171** 849–867
[14] Cardenas B, Swinfen-Styles L, Rouse J, Garvey S D. 2021 *Energies* **14** 8534
[15] Gridwatch: GB national grid status, Available online: https://www.gridwatch. templar.co.uk/
[16] Elexon: Balancing Mechanism Reporting Service. Actual Generation Output Per Generation Unit. Available online:https://www.bmreports.com/bmrs/?q=actgenration/actualgeneration. Accessed in July 2024
[17] Pfenninger S, Staffell I. Renewables Ninja. Available online: www.renewables.ninja. Accessed in July 2024
[18] The Crown Estate. Project listings: September 2023 UK offshore wind farm projects from leasing rounds through to operational phases. Available at: https://www.thecrownestate.co.uk/media/3954/offshore-wind-project-listing.pdf Accessed in July 2024
[19] The Wind Power. Vestas V164/9500. Available online at: https://www.thewindpower.net/turbine_en_1476_vestas_v164-9500.php Accessed in July 2024
[20] Abrahamsen A B, Beauson J, Lund K W, Madsen E S, Rudolph D P, Jensen J P. 2024 *Wind Energy* **27** 165-178
[21] Evans S (8 July 2022) Analysis: Record-low price for UK offshore wind is nine times cheaper than gas. Carbon Brief. Accessed in July 2024
[22] Offshore Renewable Energy Catapult (2019) Wind Farm Costs. Available online at: https://guidetoanoffshorewindfarm.com/wind-farm-costs Accessed in July 2024
[23] Hutchins M (12 December 2018). Solar 'could soon be UK's cheapest source of energy. PV Magazine
[24] Wang J, Lu K, Ma L, Wang J, Dooner M, Miao S, Li J, Wang D. 2017 *Energies* **10** 991
[25] Tarkowski R. 2019 *Renew. Sustain. Energy Rev.* **105** 86–94
[26] Blanco H, Faaij A. 2018 *Renew. Sustain. Energy Rev.* **81** 1049-1086
[27] UK Department of Energy and Climate Change (2016). Electricity Generation Costs and Hurdle Rates Lot 3: Non-Renewable Technologies pp.19.
[28] Barbour E, Mignard D, Ding Y, Li Y. 2015 *Applied Energy* **155** 804–815
[29] Victoria M, Zhu K, Brown T., Andresen G B., Greiner M. 2019 *Energy Conv. Manag.* **201** 111977
[30] Lyseng B, Niet T, English J, *et al.* 2018 *Int. J. Hydrogen Energy* **43** 1966-1979
[31] Jülch V. 2016 *Applied Energy* **183** 1594-1606

8th Offshore Energy & Storage Symposium (OSES2024) IOP Publishing
Journal of Physics: Conference Series **2929** (2024) 012007 doi:10.1088/1742-6596/2929/1/012007

Electrifying Maritime Shipping: Evaluating the CO_2 Reduction and Viability of Future Battery-Powered Container Ships Calling at the Port of Los Angeles

Jules Wilson [1], Rupp Carriveau [1*], William Hurley [2], Reza Babaei[1], David S-K Ting[1]

1 Environmental Energy Institute, University of Windsor, Windsor, Ontario, Canada
2 Hurley Maritime Consulting, Seattle, Washington, United States
*Corresponding Author: rupp@uwindsor.ca

Abstract. Maritime shipping plays a pivotal role in global logistics, influencing efficiency, emissions, and adaptability. This academic study envisions future electrified shipping fleets and evaluates the potential for CO_2 reduction and the cost delta between grid electricity and marine fossil fuels using three types of container ships (5000-7999 TEU, 8000-11999 TEU, and 12000-14499 TEU). The scenario evaluated envisions an electrical charging infrastructure at the Port of Los Angeles, supported fully by wind and solar energy sources, to recharge future battery-powered ships. It is recognized that electric-powered ocean-going long-transit ships are not practical, but this study provides baseline groundwork to justify and support future development of hybrid or partially electric ships to help meet the International Maritime Organization's (IMO) target of a 50% reduction in CO_2 emissions from the world's shipping by 2050. Additionally, a real case study analysis of all container ships at POLA was conducted using container vessel activity to further evaluate the practical implications and benefits of this transition. This analysis underscores the significant electrical infrastructure investments required and highlights the potential economic challenges as well as the environmental benefits of integrating electric power into maritime operations.

Keywords: Maritime shipping, electrification, renewable energy, CO_2 emissions

1. Introduction

The electrification of ships is gaining momentum as a sustainable solution to reduce emissions and enhance efficiency in maritime transportation. This transition begins with fully electric short-transit ferries and harbor tugboats where charging infrastructure can support frequent charging cycles and battery capacity can be limited so that battery weight and space can be accommodated in the vessel design. Incentives for electrification include significant reductions in greenhouse gas emissions, helping to meet the International Maritime Organization's (IMO) target of a 50% reduction in CO2 emissions from global shipping by 2050. Through hybridization methods, ships can leverage electric power alongside conventional fuel systems, leading to improved energy efficiency and lower operating costs [1], [2]. Additionally, electrification enables the adoption of innovative technologies such as battery storage, fuel cells, and quick-charge proximity shore power connections, thus further advancing the maritime industry toward a greener and more sustainable future. While one can assume that future technology advances will support a gradual migration of battery-power to larger ships on longer transits, this migration is not obvious and needs analysis.

The feasibility analysis of electrification initiatives is pivotal due to its role in assessing the practicality, viability, and financial implications of implementing such endeavors [3], [4]. Through a comprehensive feasibility analysis, stakeholders can meticulously evaluate diverse factors, including technical requisites, cost-effectiveness, environmental ramifications, regulatory adherence, and stakeholder reception [5], [6]. This analysis yields invaluable insights into the feasibility and long-term sustainability of the electrification initiative. Furthermore, it facilitates the identification of potential hurdles and opportunities, enabling stakeholders to make well-informed decisions and formulate strategies to surmount challenges. Ultimately, prioritizing the feasibility of electrification initiatives ensures efficient resource allocation and the attainment of desired objectives [7].

Container ships play a critical role in global trade, transporting a vast array of goods across the world's oceans with remarkable logistical efficiency. They are the backbone of the international shipping industry, enabling the seamless movement of consumer goods, industrial products, and raw materials. The standardized containers they carry facilitate easy loading and unloading, reducing turnaround times in ports, and enhancing supply chain efficiency. Container ships are designed to handle large volumes of cargo, which translates to significant energy consumption and higher levels of CO_2 emissions [8]. Specifically, at the Port of Los Angeles (POLA), one of the busiest ports in the world, container ships are vital to the region's economy, handling millions of twenty-foot equivalent unit (TEU) containers annually. The high energy demands of these ships underscore the need for innovative solutions, such as electrification, to reduce their environmental impact while maintaining their operational benefits and economic advantages. The electrification of container ships at POLA could significantly contribute to the port's sustainability goals, reducing greenhouse gas emissions and supporting California's environmental initiatives, thus enhancing the port's role as a leader in green maritime practices.

In this study, three different capacities of container ships—5000-7999 TEU, 8000-11999 TEU, and 12000-14499 TEU—each with unique performance and operational features, were introduced to evaluate their electrical demand and CO_2 emissions. The study uses data from Reference 8 on the average main engine power, average days at sea, and average fuel consumption for each ship's capacity. The feasibility of charging these ships with clean electric power at the Port of Los Angeles (POLA) during the offloading and loading operations was also investigated.

The study evaluates how electrification can reduce CO_2 emissions and support California's environmental initiatives, highlighting the potential for significant environmental benefits through innovative maritime practices.

The specific aspects of this study are:

i. **Calculation of electrical load:** The study quantifies the electrical load requirements for each ship type during their maritime operations, accounting for varying durations spent navigating open waters. This includes an analysis of the electricity demand and the capacity required to ensure uninterrupted operations, highlighting the energy needs of container ships with different TEU capacities.

ii. **Assessment of potential CO_2 emissions:** This section emphasizes the environmental impact of current maritime operations and explores the reductions achievable through electrification. An examination of potential CO_2 emissions across the three vessel types provides a comparative analysis of emissions and the potential for CO_2 reductions. Additionally, the impact of slower ship speeds on fuel consumption and CO_2 emissions is analyzed, highlighting how speed management or "slow steaming" influences emissions and fuel efficiency.

iii. **Assessment of potential cost deltas:** The potential electrical energy costs are compared to the cost of marine fossil fuels. While a full economic analysis is beyond the scope of this report, the comparison of these operational costs offers insight into the financial viability of adopting electrification technologies.

iv. **Analysis of IMO's net-zero roadmap**: The study reviews the targets outlined in the International Maritime Organization's (IMO) net-zero roadmap, assessing the feasibility and implications for maritime emissions reduction efforts from 2025 to 2050.

v. **Brief case study:** This case study analyzes the electrical requirements and potential benefits of electrifying container ships operating between the Port of Los Angeles (POLA) and Shanghai. The analysis quantifies the energy demands, examines transit times, and assesses the feasibility of supporting this transition with renewable energy sources. It provides a comprehensive overview of the practical implications and benefits of this electrification strategy, highlighting its scalability and the impact of reducing carbon emissions on one of the world's busiest shipping routes.

2. Background

Various supporting studies provide background on various aspects of reducing the environmental impact of marine operations, including the analysis of CO_2 emissions from maritime vessels, the exploration of hybrid energy systems in maritime vessels, the transition to battery-electric ships, and the integration of electrical energy storage in ships.

Anwar et al. [9] evaluated the electrification of ferries to mitigate CO_2 emissions and environmental impact. The analysis encompassed 26 commercialized vessels and 10 case studies, examining parameters such as battery capacity, power train, and vessel size. The data indicated that fully electric vessels, such as Denmark's "Ellen," possess high battery capacities (4,300 kWh) due to their pure electric power train, whereas hybrid vessels, such as the USA's "MV Waterman," have lower capacities (80 kWh) as they also rely on diesel engines. It was demonstrated that electrification significantly reduces noise pollution and emissions during docking, thereby optimizing battery use. Trends in the maritime sector towards green transformation were identified, underscoring the importance of electric and hybrid systems in achieving sustainability goals.

Daniel et al. [10] investigated the implementation of shore power, or cold ironing, to reduce emissions from ships while at berth. The study found a significant reduction in greenhouse gas emissions, with a 7.8% improvement in the Carbon Intensity Index (CII) for the analyzed vessel during loading or unloading. It advocated for including shore power in the calculations of the Energy Efficiency Design Index (EEDI) and Energy Efficiency Existing Ship Index (EEXI), emphasizing its crucial role in reducing emissions while at berth. Additionally, policy enhancements were recommended to integrate shore power into routine shipping operations, highlighting its immediate environmental benefits and potential to advance eco-friendly shipping practices. Cold-ironing shore power infrastructure supports only auxiliary power loads, and significant shore-side upgrades would be needed to support battery charging for propulsion power during transit.

Gillingham and Huang [11] evaluated the implications of electrification strategies within the U.S. waterborne shipping industry through various scenarios using the Yale-NEMS model. It revealed that electrification, especially when combined with carbon pricing policies, could significantly decrease CO_2 and local air pollutant emissions, suggesting a potential net reduction of up to 65% by 2050. Additionally, the study found that while initial costs for infrastructure and retrofitting were substantial, the long-term social and environmental benefits of reduced pollution and improved public health significantly outweighed these costs. The research emphasized the importance of a cleaner electricity grid to maximize the benefits of shipping electrification and suggested that strategic policy frameworks were essential to facilitate the transition towards sustainable maritime transport.

Park and Hendrickson [12] examined the transition to battery-electric ships (BES) in the U.S. maritime sector. It was found that electrifying 6,323 domestic ships could reduce greenhouse gas emissions by up to 75% by 2035, with 85% of these ships achieving cost-effectiveness compared to internal combustion ships if they operate on a decarbonized grid. Challenges such as the scale of battery production and the physical capacity of ships to accommodate increased battery weights were addressed. The study proposed that with strategic planning, the number of ports that would need significant infrastructure upgrades to handle the new power demands could be limited. This research supports the viability of BES as a sustainable alternative within the domestic shipping industry, highlighting the critical need for infrastructure adaptation at concentrated locations.

Helgesson [13]conducted a feasibility study on the decarbonization and electrification of a commuter ferry in Stockholm, evaluating Internal Combustion Engine (ICE) Diesel, ICE with Hydrotreated Vegetable Oil (HVO), Li-ion battery, and fuel cell hybrid systems. The study found that the ICE Diesel system had the lowest capital cost but the highest replacement cost and CO_2 emissions (2.97 kg CO_2/L). The ICE HVO system exhibited moderate emissions (0.70 kg CO_2/L) with a similar capital cost. The Li-ion battery system, despite a high capital cost of $837,648, achieved the lowest variable emissions (0.013 kg CO_2/kWh) due to Sweden's renewable-heavy grid, though it faced challenges related to weight and charging requirements (1,939 kWh battery capacity). The fuel cell hybrid system, with a capital cost of $494.7/kW for power electronics, showed lower emissions dependent on hydrogen production (2.0 kg CO_2/kg H_2 from electrolysis) but necessitated significant infrastructure changes. The study concluded that HVO and fuel cell hybrids are the most viable options, offering a balance of environmental benefits and practical constraints.

3. Methods and Approach

3.1 Analyzed Ship Types

This investigation analyzed three distinct capacities of container ships, specifically categorized by their cargo capacity in TEU. These container ship sizes are the most common sizes used in the worldwide trade, each with unique performance and operational features. The analyzed ship types encompass:

1. **Container ship 5000-7999 TEU**: Designed to transport large loads of standardized containers, suitable for extensive routes and major ports. These ships have substantial cargo capacity, featuring high container stacks, a robust bridge, and powerful engines to handle significant cargo volumes and long-distance travel.
2. **Container ship 8000-11999 TEU**: Capable of carrying very large-sized loads, offering a significant balance between cargo capacity and operational efficiency. These ships are larger than the 5000-7999 TEU category, with even more container stacks, a more extensive deck area, a larger bridge, and advanced infrastructure to handle increased cargo loads and improve operational flexibility.

3. **Container ship 12000-14499 TEU**: Built to transport the largest volumes of standardized containers, ideal for the longest routes and the largest ports, maximizing cargo efficiency. These ships feature the highest container stacks, a vast bridge, and the most advanced loading/unloading equipment to handle the most significant cargo volumes and ensure efficient operations.

Table 1 provides a comparative overview of three sizes of container ships categorized by TEU capacity. It highlights their average deadweight tonnage (DWT), engine power, speed, days at sea, and fuel consumption for main engines, auxiliary engines, and boilers, for one ship. The data shows that as the ship size increases, there is a corresponding rise in DWT, engine power, speed, and fuel consumption, reflecting the greater capacity and operational demands of larger vessels.

3.2. Lithium Battery Specifications

Table 1. Comparative overview of container ship capacities

Ship type/size	Avg. DWT2 (tonnes)	Avg. engine power (kW)	Avg. speed (kt)	Avg. annual days at sea	Avg. annual consumption (kilotonne)		
					Main	Aux.	Boiler
Container/5000-7999 TEU1	74,661	52,566	24.6	258	20.3	2.4	0.5
Container/8000-11999 TEU1	110,782	57,901	23.9	261	26.4	2.9	0.5
Container/12000-14499 TEU1	149,023	61,231	23.8	246	27.2	3.3	0.6

[a1] Twenty-Foot Equivalent Unit.
[2]Average Deadweight Tonnage

This study considered the use of lithium batteries for the electrification of container ships. The specifications of the batteries used are detailed in Table 2. These specifications were chosen to ensure optimal performance and longevity, allowing for efficient energy storage and usage. The high efficiency (97%) minimizes energy loss during charging and discharging, maximizing usable energy. The initial state of charge (SoC) at 100% provides maximum operational time, while the minimum SoC at 10% preserves battery health and longevity. A depth of discharge (DoD) of 80% ensures sufficient energy availability for ship operations at sea, which is crucial for maintaining continuous power supply for main engines, auxiliary systems, and onboard equipment during long voyages without relying on fossil fuels.

There are currently available 2 MW-hr independent battery systems housed in 40-foot containers that could be loaded on the container ship with other containers and connected to the ship's electrical system. The more cost-effective approach however would be to provide a permanent space to house the batteries.

3.3. CO_2 Calculation

Table 2. Specifications of lithium batteries used in the study

Specification	Value
Nominal voltage (V)	720
Nominal capacity (kWh)	1
Lifetime (years)	20
Battery Efficiency (%)	97%
Initial State of Charge (SoC)	100%
Minimum State of Charge (SoC)	20%
Depth of Discharge (DoD) (%)	80%

Accurately calculating CO_2 emissions is essential for assessing the environmental impact of maritime operations and the potential benefits of integrating electrification. This section outlines the methodology used to calculate the annual CO_2 emissions for container ships based on their power output and the emission factors associated with their current fuel type, which is Heavy Fuel Oil (HFO). Table 3 provides details of the carbon content and emission factors for various fuel types.

Table 3. Comparative overview of container ship fuel types

Fuel type	Carbon content	EF (g CO_2/g fuel)
HFO (Heavy Fuel Oil)	0.8493	3.114
MDO (Marine Diesel Oil)	0.8744	3.206
LNG (Liquefied Natural Gas)	0.75	2.75
Methanol	0.375	1.375
LSHFO 1.0% (Low Sulphur Heavy Fuel Oil)	0.8493	3.114

$$EM_i = EF \times W_i \tag{1}$$

Where EF is the Emission Factor for CO_2, expressed in tonnes of CO_2 per tonne of fuel, and W_i is the Power Output, expressed in kilowatt-hours per year (kWh/year)

3.4. *Speed Management and Fuel Consumption*

Speed management involves reducing a vessel's speed to reduce fuel consumption and emissions. The relationship between a vessel's speed and fuel consumption is exponential, meaning small speed changes can lead to significant changes in fuel consumption.

$$P_{reduced} = P_{initial} \times \left(\frac{V_{reduced}}{V_{initial}}\right)^3 \tag{2}$$

Where P is the engine power required (proportional to fuel consumption) and V is the vessel speed

3.5. *Port of Los Angeles (POLA)*

The Port of Los Angeles, as shown in Fig. 1, serves as a vital hub for international trade and commerce. This port handles a diverse array of ship types, but container ships dominate in terms of both the number of vessels and total cargo value. Container shipping is particularly crucial for these ports, as it facilitates the seamless movement of a vast range of goods, supporting both local and global supply chains. Beyond their logistical importance, the Port of Los Angeles and the adjacent Port of Long Beach are increasingly prioritizing sustainability initiatives by adopting cleaner energy sources and advanced technologies to minimize environmental impacts. This focus on sustainability aligns with California's strong commitment to environmental stewardship and economic vitality, supporting the integration of renewable energy sources such as wind and solar power into port operations. These initiatives not only help reduce the carbon footprint of maritime activities but also promote the use of green technologies in the logistics and transportation sectors. The combined efforts of these ports and California's progressive environmental policies create a dynamic framework that supports the adoption of sustainable practices in one of the world's busiest and most significant maritime hubs, demonstrating how economic growth can be achieved in harmony with environmental conservation.

Fig. 1. Geographical location of Port of Los Angeles [14]

3.6. *International Maritime Organization's Net-Zero Roadmap*

The International Maritime Organization (IMO) plays a critical role in establishing regulations aimed at reducing greenhouse gas (GHG) emissions in the global maritime sector. Since January 1, 2000, the IMO has mandated reductions in nitrogen oxide emissions from marine diesel engines, with Tier 3 limits enforced in designated special areas and Tier 2 limits elsewhere for vessels constructed from January 1, 2011, onwards. More stringent measures were introduced on January 1, 2020, to curb sulfur oxides emissions from marine fuels, resulting in a substantial reduction—approximately sixfold—of these emissions into the atmosphere. Furthermore, the IMO Regulation on Energy Efficiency for Ships, which took effect on January 1, 2013, specifically targets the reduction of carbon dioxide emissions. In its commitment to mitigating the impact of maritime emissions on climate change, the IMO has set ambitious targets to decrease carbon dioxide emissions by a minimum of 30% by 2030 and 70% by 2050 compared to 2008 levels, with an overarching goal of reducing the equivalent carbon dioxide emissions by 50% by 2050. These measures reflect the organization's proactive stance on environmental protection within the maritime industry [15].

3.7. *Electricity Purchase Rate*

Fig. 2 displays the monthly electricity purchase rate ($/kWh) at the Port of Los Angeles (POLA) over the course of a year, extracted from California Electric Utility Retail Price [16]. The data highlights a general upward trend in purchase rates during the summer months, with the highest rates observed in August ($0.277/kWh) and July ($0.266/kWh). Conversely, the purchase rates are relatively lower in the winter and early spring months, with the lowest rate in February ($0.225/kWh). The rates gradually increase from January ($0.229/kWh) to a peak in August, before decreasing slightly towards the end of the year, with December's rate at $0.242/kWh. This seasonal variation likely reflects the increased demand for electricity during the summer months, leading to higher costs.

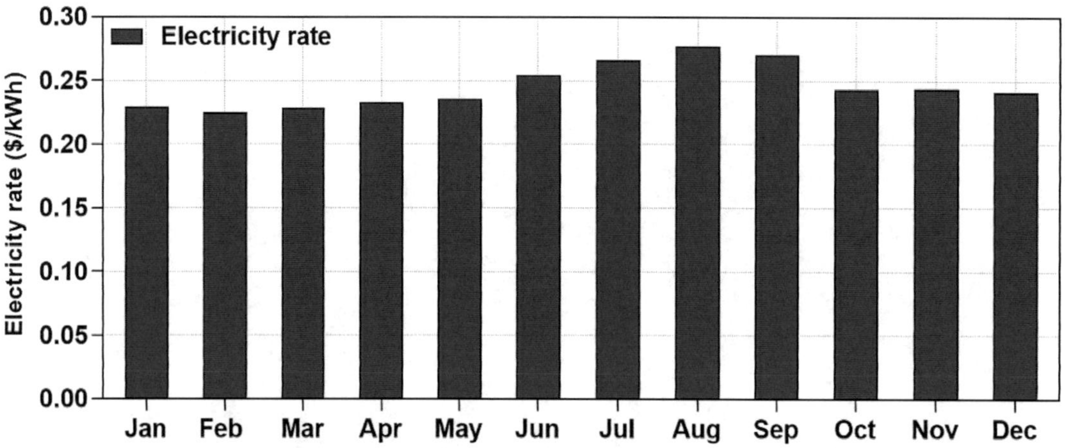

Fig. 2. Monthly electricity purchase rate of Port of Los Angeles (POLA)

4. Calculation of electrical load & fuel consumption

Accurate load calculation is essential for understanding the energy demands associated with electrifying container ships and ensuring the feasibility of this transition. By utilizing average load consumption data from the IMO report, we have estimated the total energy requirements for ship operations. This information is critical for planning the necessary capacity and infrastructure upgrades to support battery charging from the grid system, as well as for identifying the required renewable energy sources, such as wind and solar power. These calculations are vital for achieving sustainability goals and promoting eco-friendly maritime operations at the Port of Los Angeles (POLA). The calculated load demand represents the amount of energy storage needed for the ships to operate entirely on battery power for their entire one-way voyage at sea without needing to recharge. This battery capacity ensures that the ships can stay at sea for their average number of days and perform their propulsion and auxiliary functions before reaching their destination port for recharging. It becomes clear from these numbers, and from the case study presented later, that a 100% battery-electric ship (BES) is not feasible, but the analyses provide useful benchmarks for understanding the challenges. A hybrid approach integrating some amount of battery energy with other low-carbon energy sources will support the transition to sustainable maritime transport, reduce greenhouse gas emissions, and align with global sustainability goals.

Table 4 and Fig. 3 present the annual energy consumption for fully charging a 100% BES container ship at the Port of Los Angeles. When slow steaming at a reduced speed, power applied for propulsion plus the auxiliary power is about 70% of the installed power, hence this applied power is used to estimate the energy expended.

Table 4. Energy Consumption for a 100% BES container ship

Ship type	Installed Power (kW)	Applied Power at reduced speed (kW)	Average Annual Days at Sea	Annual Energy Consumption (MW-hr)
Container/5000-7999 TEU	52,566	36,796	258	227,842
Container/8000-11999 TEU	57,901	40,531	261	253,884
Container/12000-14499 TEU	61,231	42,862	246	253,055

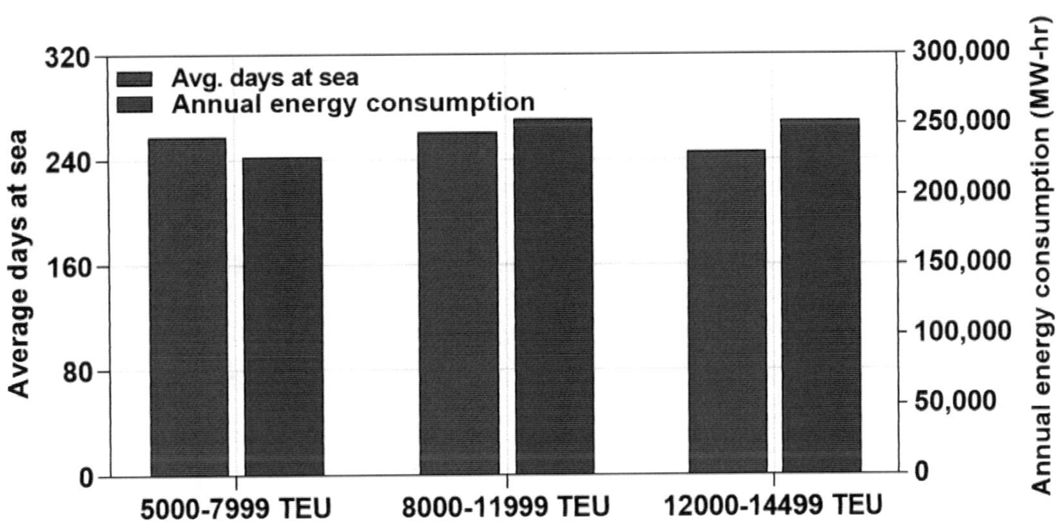

Fig. 3. Average days at sea and energy consumption for the considered container ships

Fuel consumption and associated costs are critical factors in the economic and environmental performance of container ships. By analyzing the total fuel consumption and annual fuel costs for different ship types, stakeholders can better evaluate the potential cost impacts in cash flow modeling and make informed decisions about investments in cleaner technologies. Fig. 4 illustrates the total fuel consumption and annual fuel cost for three types of container ships categorized by their TEU capacities. Container ships with capacities of 5000-7999 TEU burn 23,200 tonnes of fuel annually, resulting in a fuel cost of approximately $15.08 million. Ships in the 8000-11999 TEU category burn 29,800 tonnes of fuel annually, with a corresponding fuel cost of $19.37 million per year. The largest ships, in the 12000-14499 TEU range, burn 31,100 tonnes of fuel annually, incurring a fuel cost of around $20.22 million.

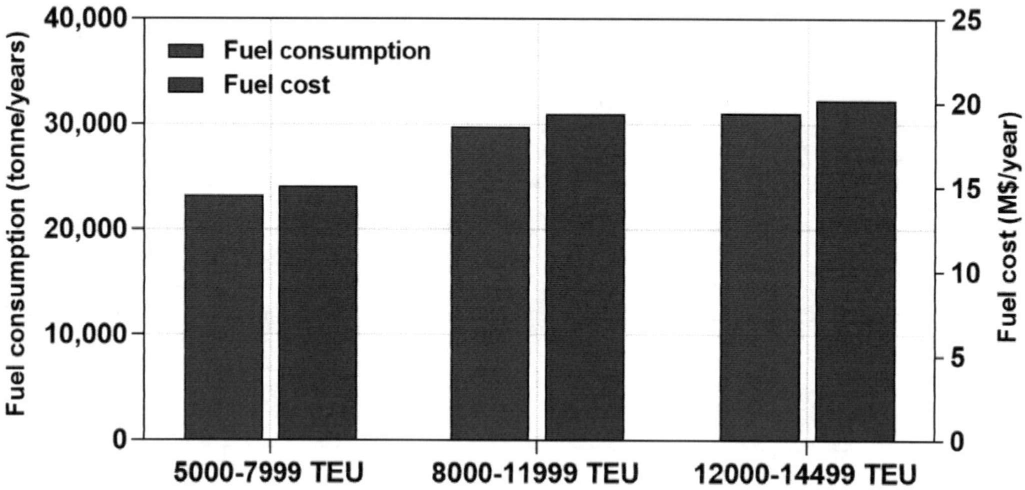

Fig. 4. Total fuel consumption and annual fuel cost for the considered container ships

5. Assessment of potential CO_2 reduction

Analyzing CO_2 reduction is essential for understanding the environmental impact of container ship operations. Larger ships with higher fuel consumption emit greater amounts of CO_2, exacerbating their environmental footprint. The International Maritime Organization (IMO) has set a target to reduce CO_2 emissions by 50% by 2050, underscoring the necessity for substantial changes within the maritime industry.

Fig. 5 displays potential CO_2 reduction, per ship, for the three container ship categories. These values are the estimates of what the ship's current emissions are. Container ships in the 5000-7999 TEU range achieve a CO_2 reduction of 72,245 tonnes/year. The 8000-11999 TEU ships show a CO_2 reduction of 92,797 tonnes/year. The 12000-14499 TEU ships demonstrate a CO_2 reduction of 96,845 tonnes/year. This data highlights the significant environmental benefits of electrification, with larger ships showing greater CO_2 reductions.

Another comparison that measures emissions per quantity of goods transported is CO_2 emission per TEU-mile. By this measure the larger the ship the lower the CO_2 emission per quantity of goods transported. This comparison also applies to fuel cost, which is why ship owners prefer to build and operate larger ships. CO_2 (grams)/TEU-mile values are 100 for the small category, 82 for the mid-category, and 68 for the large ship category.

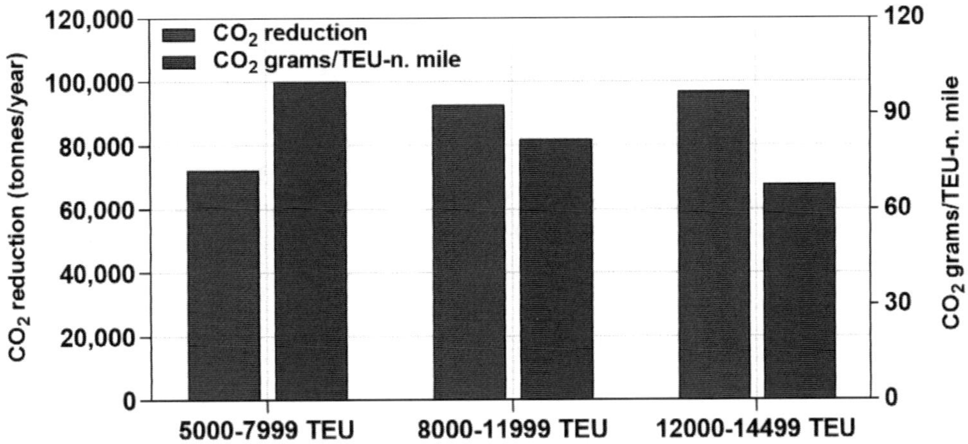

Fig. 5. Total CO_2 reduction and emissions per TEU-mile for the considered container ships

Understanding the impact of ship speed on fuel consumption and CO_2 emissions is crucial for optimizing maritime operations and minimizing environmental impacts. Higher ship speeds increase hydrodynamic resistance, leading to exponential rises in fuel consumption and associated CO_2 emissions. By analyzing this correlation, stakeholders can develop strategies for fuel efficiency, cost reduction, and environmental compliance, aligning with sustainability goals and international regulations set by bodies such as the International Maritime Organization (IMO).

The reduction of ship speeds to lower emissions and fuel consumption has been extensively discussed and analyzed over the years. According to the IMO report, container ships in the 8000-11999 TEU range are already operating at a reduced speed of 16 knots, down from their design speed of approximately 22 knots. Speed reductions have been implemented on most ships worldwide, driven more by economics than emission reduction, and further reductions may not be practical or feasible.

Table 5 presents the impact of varying ship speeds on fuel consumption and CO_2 emissions for three types of container ships. At baseline speeds of 24.6, 23.9, and 23.8 knots respectively,

Table 5. Fuel Consumption and Emissions for a 100% BES container ship

Ship Type	Speed (knots)	Reduced power (kW)	Fuel consumption (tonnes/year)	CO_2 emissions (tonnes/year)
Container 5000-7999 TEU	18.45 (25% lower)	14,142	21,650	67,689
	24.6 (Baseline)	52,566	23,200	72,245
	30.75 (25% higher)	92,791	29,000	90,306
Container 8000-11999 TEU	17.93 (25% lower)	15,308	27,150	84,769
	23.9 (Baseline)	57,901	29,800	92,797
	29.88 (25% higher)	99,537	38,500	119,683
Container 12000-14499 TEU	17.85 (25% lower)	16,204	28,533	88,844
	23.8 (Baseline)	61,231	31,100	96,845
	29.75 (25% higher)	108,453	40,400	125,788

the annual fuel consumption and CO_2 emissions are 23,200 tonnes/72,245 tonnes, 29,800 tonnes/92,797 tonnes, and 31,100 tonnes/96,845 tonnes. Reducing speed by 25% significantly lowers fuel consumption and emissions, while increasing speed by 25% results in substantial increases in both metrics. For example, a 5000-7999 TEU ship at 18.45 knots consumes 21,650 tonnes of fuel and emits 67,689 tonnes of CO_2, whereas at 30.75 knots, it would consume 29,000 tonnes of fuel and emit 90,306 tonnes of CO_2. This analysis underscores the critical role of speed management in minimizing the environmental footprint of maritime operations while acknowledging the practical limitations of further speed reductions.

6. Assessment of potential cost deltas

Using the current cost of Heavy Fuel Oil (HFO) and the current cost of electricity in California, the operational economics from simply paying for California electricity instead of fossil fuel strongly favor fossil fuel. Capital costs, while not analyzed here, will similarly favor fossil fueled shipping. The underlying challenge in this case is the cost of electricity in California, currently about \$0.24/kW-hr, which results in an annual electricity cost of \$54.8M for an 8,000 TEU container ship. Current fuel cost of \$650/tonne results in an annual fuel cost of \$17.2M, about a third of the electricity cost. It is important to note that if California electricity costs can get down to \$0.08/kW-hr, costs that are prevalent in other parts of the country, the cost comparison would be even. If fuel cost increased to \$850/ton the break-even electricity cost would be \$0.10.

7. Analysis of IMO net-zero roadmap

IMO's 50% emissions reduction target aligns with the ambitions of several nations, including Japan, the UK, and the USA, which have committed to achieving net-zero emissions in maritime transport by the same year. Consequently, the shipping industry is under significant pressure to adopt zero-emission technologies within the next decade to meet these goals. Although the necessary technologies exist, they must be adopted more broadly, quickly, and cost-effectively. The adoption of zero-emission fuels, which are significantly more expensive than traditional fuels, could increase the total ownership costs of vessels by 40% to 60%, varying by shipping route.

Fig. 6 illustrates the proposed CO_2 reduction plan for different container ship types from 2025 to the IMO's net-zero target year of 2050. If only a 50% reduction in emissions is planned to be implemented by 2050, regardless of fully electrifying these ship types, the emissions trajectory is outlined as follows. Initially, in 2025, the combined emissions from all ship types are significant, but this total decreases gradually each year, reflecting the industry's adoption of greener technologies and fuels. By 2030, emissions have reduced significantly, and the decline will continue steadily through 2040. The rate of reduction accelerates in the final decade towards 2050, culminating in a dramatic drop, indicating a robust implementation of zero-emission technologies across the fleet. This systematic decline in emissions from 2025 to 2050 demonstrates the profound impact of adopting clean electrification technologies across maritime operations.

Journal of Physics: Conference Series **2929** (2024) 012007 doi:10.1088/1742-6596/2929/1/012007

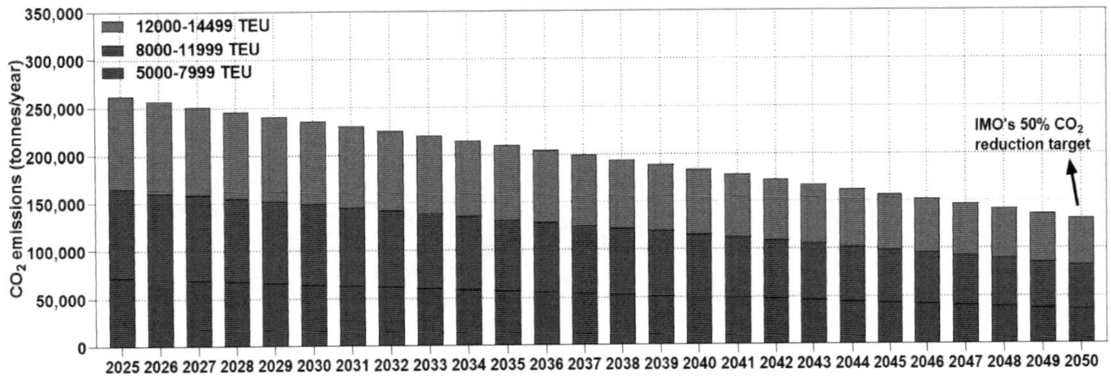

Fig. 6. Maritime shipping roadmap to 50% CO_2 emissions reduction by 2050

8. Case study of container ships from Port of Los Angles to Shanghai

This case study analyzes the electrical requirements and potential benefits of electrifying container ships operating between the Port of Los Angeles (POLA) and Shanghai, quantifying the associated infrastructure upgrades to meet the electrical supply requirements. By examining energy consumption, transit times, and overall feasibility, this study highlights the potential for reducing carbon emissions and enhancing the sustainability of maritime operations on one of the world's busiest shipping routes.

Fig. 7 illustrates the maritime shipping route from the Port of Los Angeles (POLA) in the United States to Shanghai, China. This visual representation emphasizes the substantial distance and the strategic importance of this route for international trade. As a key segment in global logistics, this route connects two of the world's largest and busiest ports.

This analysis calculates the electrical requirements for recharging container ships transiting from POLA to Shanghai. Referencing data available from the POLA, the average number of imported containers to the port annually is about 5,000,000 TEUs. Simplifying the analysis by assuming all ships were of 8,000 TEU capacity results in 625 container ship calls per year, about two ships a day. Further simplifying, this quantity of ship calls could be made by 51 - 8,000 TEU ships working full time for a year, running back and forth with 2 days for operations at each end.

This case study illustrates the large magnitude of the electricity that would be consumed by recharging all the container ships calling at POLA, assuming they were all fully electric battery-energy ships. From this upper bound and recognizing the technical and economic constraints, work can be done to scale-back to a feasible and executable quantity of electricity for integration into the ship's systems, and a feasible amperage of electricity that can be delivered to the berths at POLA. For perspective, providing this 100% case would require a renewable generation capacity of 1.5GW, assuming a new offshore wind farm with a 50% capacity factor. It also implies that 18 GW-hr of power be provided by the Port daily. For reference, California's total clean power capacity is currently approximately 30,000 MW [17].

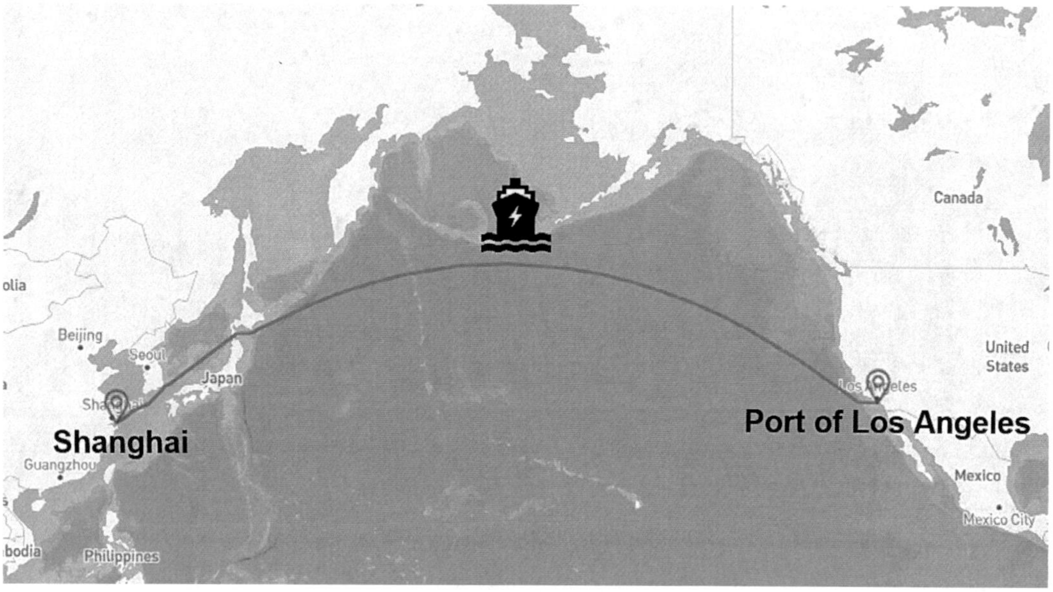

Fig. 7. Maritime shipping path from POLA to Shanghai

9. Conclusions

This study provides an analysis of the potential benefits and challenges of electrifying three types of container ships: 5000-7999 TEU, 8000-11999 TEU, and 12000-14499 TEU, using clean grid energy from the Port of Los Angeles (POLA). The purpose of electrifying these ships is to reduce CO_2 emissions, optimize operational efficiency, and support sustainability goals. The key findings or take-aways from the results and discussion sections are summarized as follows:

i. Container ships demand substantial power and energy. Annual energy requirements per ship average about 250 GW-hr. To support this energy demand about 28 MW of dedicated power is required, per ship, equivalent to 4-15MW wind turbines operating at about a 50% power factor.

ii. Significant reductions in CO_2 emissions could be realized by electrifying ships with renewable energy. CO2 emissions for the three ship sizes investigated are reduced by 72,245 tonnes, 92,797 tonnes, and 96,845 tonnes respectively, per year per ship. Managing ship speeds can also significantly impact fuel consumption and emissions, underscoring the importance of operational strategies.

iii. For the electrical energy cost to be equal to the current heavy fuel oil cost the cost of electricity in California would need to be reduced from the current \$0.24 / kW-hr to \$0.08 /kW-hr. Alternatively, a 30% increase in heavy fuel cost would be cost-equivalent to a \$0.10 /kW-hr electricity price. Levelized Cost of Energy and Cost of Transportation studies would most likely show a higher cost for electrification, as capital expenses for integrating any level of electric storage capacity are high.

iv. The case study focusing on 8,000 TEU ships calling at the Port of Los Angeles and transiting to Shanghai demonstrates that a fully electric future, with all container ships calling at POLA being fully recharged, does not currently appear feasible. The

Port would need to deliver 18 GW-hr of clean energy daily and would need a 1.5 GW offshore wind farm to support it. However, establishing this upper bound is useful in evaluating lower amounts of electric power integration that when combined with other low CO_2 energy fuels and sources can help attain the IMO goals.

10. Closing

The analysis data and information presented in this paper are intended to support the case for at least partial electrification of the future shipping fleet. There are other related questions that economic analysis needs to address, such as ship modification costs (or cost deltas in newbuild ships), and infrastructure costs to provide electric power capacity at the ports. There is also the question of how much battery power can be practically integrated into the design and function of a large ship, and how this battery integration compares with the other low-carbon fuels and propulsion-assist devices currently being explored [18]. Available renewable electricity and its cost in the future is unknown, as there is competing demand for renewable energy in our terrestrial infrastructure needed to reduce those many emissions.

The shipping industry is currently focused on alternative fuels such as methanol, as migrating to these fuels is easier to integrate into current ship design and ship propulsion engines. This is a good first step, and future steps could include electric power integration into the shipping industry.

Acknowledgments

We express our sincere gratitude to Glosten, Naval Architecture and Marine Engineering, headquartered in Seattle, Washington, USA, and their marine engineer Colin Flynn, for their invaluable contributions to this study. Their expertise and support have been instrumental in providing perspectives that enriched the quality and depth of our research.

References

[1] X. Liu *et al.*, "Techno-economic analysis of solar tracker-based hybrid energy systems in a rural residential building : A case study in South Africa," *Int. J. Green Energy*, vol. 00, no. 00, pp. 1–20, 2022, doi: 10.1080/15435075.2021.2024545.

[2] R. Babaei, D. S. Ting, and R. Carriveau, "ScienceDirect Optimization of hydrogen-producing sustainable island microgrids," *Int. J. Hydrogen Energy*, vol. 47, no. 32, pp. 14375–14392, 2022, doi: 10.1016/j.ijhydene.2022.02.187.

[3] A. Amirsoleymani, R. Babaei, S. S. Mousavi Ajarostaghi, and M. Saffari Pour, "Feasibility evaluation of Stand-Alone energy solutions in Energy-Poor Islands using sustainable hydrogen production," *Int. J. Energy Res.*, vol. 46, no. 15, pp. 24045–24063, 2022, doi: 10.1002/er.8704.

[4] Q. Ma, X. Huang, F. Wang, C. Xu, R. Babaei, and H. Ahmadian, "Optimal sizing and feasibility analysis of grid-isolated renewable hybrid microgrids: Effects of energy management controllers," *Energy*, p. 122503, Nov. 2021, doi: 10.1016/j.energy.2021.122503.

[5] R. Babaei, D. S. Ting, and R. Carriveau, "Feasibility and optimal sizing analysis of stand-alone hybrid energy systems coupled with various battery technologies : A case study of Pelee Island," *Energy Reports*, vol. 8, pp. 4747–4762, 2022, doi: 10.1016/j.egyr.2022.03.133.

[6] T. Chen, M. Wang, R. Babaei, M. E. Safa, and A. A. Shojaei, "Technoeconomic Analysis and Optimization of Hybrid Solar-Wind-Hydrodiesel Renewable Energy Systems Using Two Dispatch Strategies," *Int. J. Photoenergy*, vol. 2023, 2023, doi: 10.1155/2023/3101876.

[7] H. Kim, K. Y. Koo, and T.-H. Joung, "A study on the necessity of integrated evaluation of alternative marine fuels," *J. Int. Marit. Safety, Environ. Aff. Shipp.*, vol. 4, no. 2, pp. 26–31, Apr. 2020, doi:

10.1080/25725084.2020.1779426.
[8] IMO, "Fourth IMO GHG Study 2020 Full Report," *Int. Marit. Organ.*, vol. 6, no. 11, pp. 951–952, 2021.
[9] S. Anwar, M. Y. I. Zia, M. Rashid, G. Z. De Rubens, and P. Enevoldsen, "Towards ferry electrification in the maritime sector," *Energies*, vol. 13, no. 24, pp. 1–22, 2020, doi: 10.3390/en13246506.
[10] H. Daniel, J. P. F. Trovão, and D. Williams, "Shore power as a first step toward shipping decarbonization and related policy impact on a dry bulk cargo carrier," *eTransportation*, vol. 11, no. April 2018, 2022, doi: 10.1016/j.etran.2021.100150.
[11] K. T. Gillingham and P. Huang, "Long-run environmental and economic impacts of electrifying waterborne shipping in the United States," *Environ. Sci. Technol.*, vol. 54, no. 16, pp. 9824–9833, 2020, doi: 10.1021/acs.est.0c03298.
[12] W. Y. Park and T. Hendrickson, "Exploring the cost and emissions impacts , feasibility , and scalability of battery-electric ships," pp. 1–22, 2024.
[13] M. Helgesson, "Feasibility study of the decarbonisation and electrification of a commuter ferry in Stockholm city," KTH, School of Engineering Sciences in Chemistry, Biotechnology and Health (CBH), 2020. [Online]. Available: https://kth.diva-portal.org/smash/record.jsf?pid=diva2%3A1473009&dswid=7875
[14] E. F. Merchant, "Can Biden's Plan to Boost Offshore Wind Spread West?," *https://insideclimatenews.org/Inside Climate News*, 2021. [Online]. Available: https://insideclimatenews.org/news/26052021/biden-newsom-california-offshore-wind/
[15] A. S. Grzelakowski, J. Herdzik, and S. Skiba, "Maritime Shipping Decarbonization: Roadmap to Meet Zero-Emission Target in Shipping as a Link in the Global Supply Chains," *Energies*, vol. 15, no. 17, 2022, doi: 10.3390/en15176150.
[16] California Electric Utility Retail Price (I:CEURP), "California Electric Utility Retail Price (I:CEURP)." [Online]. Available: https://ycharts.com/indicators/california_electric_utility_retail_price
[17] Governor Gavin Newsom, "Governor Newsom Updates the Roadmap to California's Clean Energy Future." [Online]. Available: https://www.gov.ca.gov/2023/05/25/governor-newsom-updates-the-roadmap-to-californias-clean-energy-future/
[18] Globalialogisticsnetwork, "How the container shipping industry is embracing alternative fuels to boost its sustainability endeavours." [Online]. Available: https://www.globalialogisticsnetwork.com/blog/2024/06/12/how-the-container-shipping-industry-is-embracing-alternative-fuels-to-boost-its-sustainability-endeavours/

8th Offshore Energy & Storage Symposium (OSES2024) IOP Publishing
Journal of Physics: Conference Series **2929** (2024) 012008 doi:10.1088/1742-6596/2929/1/012008

Moving Beyond Diesel Generators: Exploring Renewable Backup Alternatives for Data Centers

Viswambher Kambhampati[1], Andy van den Dobbelsteen[1, 2] and Joep Schild[3]

[1] Sustainability Co-ordination, TU Delft, Delft, The Netherlands
[2] Faculty of Architecture and the Built Environment, TU Delft, Delft, The Netherlands
[3]Information and Communication Technology, Campus Real Estate & Facility Management, TU Delft, Delft, The Netherlands

E-mail: vis.kambhampati@gmail.com

Abstract. This study investigates sustainable alternatives to diesel generators for data centre backup power, focusing on renewable diesel (HVO), Hydrogen energy storage (HES), batteries (Lithium-ion and Sodium Sulfur) and Compressed Air Energy Storage (CAES). As environmental scrutiny of data centres grows, the need for cleaner energy sources intensifies. Our research assesses various storage technologies' energy performance metrics, environmental impacts, and economic feasibility. HVO is a seamless substitute for conventional diesel, compatible with existing infrastructure and less carbon-intensive. CAES offers lower life cycle emissions and operational costs but is geographically dependent. While currently more costly, batteries could achieve better economics with increased operational hours. However, extending the backup duration increases their capital and operating costs significantly, which is less advantageous than other technologies, where only fuel costs increase with longer backup times. For existing data centres transitioning to sustainable energy, HVO is optimal; for new facilities, CAES is ideal if geography allows, with HES as a robust alternative. This analysis offers a pathway for data centres to adopt sustainable, cost-effective energy storage solutions and reduce carbon footprints through on-site renewables or green energy procurement.

1 Introduction

Data centres have become the backbone of the modern digital landscape, pivotal in the rapid advancements of artificial intelligence(AI) and machine learning. As the world embraces the AI revolution, the importance of data centres becomes increasingly evident, serving as the foundational infrastructure that powers the data-driven innovations transforming various industries [1].

(a) (b) (c)

Figure 1: (a) Meta's datacentre [2] (b) Server room interior in data centre [3] (c) TU Delft datacentre

Data centres, categorized by power capacity, address various enterprise needs, as shown in Table 1. Operating 24/7 throughout the year, they are characterized by high energy intensity, with power densities typically ranging from $538\,\mathrm{W\,m^{-2}}$ to $2153\,\mathrm{W\,m^{-2}}$, and occasionally reaching up to $10\,\mathrm{kW\,m^{-2}}$ [4], [5]. In 2018, data centres worldwide consumed 205 TW h, exceeding the annual consumption of countries like Ireland and Denmark [6], [7]. By 2030, data centres are projected to consume approximately 3–13% of the world's electricity, a significant increase from the 1% consumed in 2010 [8], [9]. They are responsible for 2.5% to 3.7% of global greenhouse gas emissions, surpassing those from the aviation industry (2.4%) [10]. The global data centre market is estimated at $229 billion in 2023 and is projected to reach $641 billion by 2032 [11].

Data Centre Type	Power Usage	Description
Server room	<50kW	Supports business activities within or separate from buildings
Very small data centre	50-250kW	
Small data centre	250-1000kW	Critical for business functionality, can be in or outside buildings
Medium size data centre	1-2MW	Vital infrastructure for various businesses located in separate buildings
Large data centre	2-10MW	
Very large data centre	>10MW	

Table 1: Classification of data centres based on size [12]

Imagine a day without reliable access to essential online services like cloud storage and communication platforms. Such a disruption would significantly affect daily life. This scenario highlights the critical role of data centre reliability in maintaining uninterrupted access to the services we depend on for communication, entertainment, and productivity. Diesel generators are crucial for ensuring this reliability by providing backup power. Their use as backup systems is due to their established technology, widespread availability, high energy density, and rapid response times. However, relying on these generators increases greenhouse gas emissions due to the presence of sulfur and aromatic compounds.

Extensive research into data centre load variability and composition is crucial for designing effective backup power systems and improving energy efficiency. Studies have investigated power consumption distribution across data centre components, particularly regarding HVAC requirements based on geographic location. Shehabi et al. [13] found that server power consumption accounts for approximately 43% of total electricity use in U.S. data centres, with cooling and power provisioning systems consuming about 23% and 11%, respectively. The study also noted significant geographical variations, indicating that cooler climates require less energy for cooling. Similarly, Dayarathna et al. [14] reported that IT equipment (servers, storage, and network) typically comprises 55-60% of total power usage, while cooling systems consume 30-35%, and power distribution losses account for 10-15%. These findings underscore the importance of understanding load dynamics for developing effective backup power solutions. Another paper by Ghatikar et al. [15] examined the load flexibility of data centres and their potential for demand response, emphasizing the importance of rapid response times in backup power systems.

Regulatory bodies and the public are increasingly scrutinizing the environmental impact of data centres. As a result, technology firms are investing in renewable energy production to offset the energy consumption of data centres [16], [17], [18]. Amazon, Meta, and Google have collectively committed to generating 22 GW of renewable energy to meet their Net Zero targets. To achieve sustainability goals, traditional backup power sources like diesel generators must be replaced with reliable, widely available, and mature sustainable alternatives.

eBay's innovative data centre, operational since 2013 in South Jordan, Utah, has replaced diesel generators with Bloom Energy fuel cells powered by natural gas, thereby reducing emissions and enhancing reliability [19], [20]. NorthC is utilizing green Hydrogen for emergency power at its Groningen facility and plans to implement hybrid generators that can operate on both natural gas and Hydrogen at its upcoming site in Eindhoven, Netherlands [21]. Microsoft Sweden's data centres are leading the adoption of Preem Evolution Diesel Plus, a low-carbon fuel composed of over 50% renewables, significantly reducing net CO_2 emissions [22]. In New York, they have trialled a 3-MW hydrogen-powered system for zero-carbon data centre backup [23].

While numerous studies have explored alternatives to diesel generators for various applications, there is a noticeable gap in comprehensive comparative analyses that integrate energy performance, environmental impact assessment, and financial analysis. Existing research includes evaluations of medium-speed diesel generator sets and energy storage technologies to reduce fuel consumption and exhaust emissions in electric propulsion systems for platform supply vessels (PSVs) [24]. Other studies have focused on sustainable electricity generation using solar PV/diesel hybrid systems without storage for off-grid areas, as well as the optimization of off-grid photovoltaic-diesel-battery hybrid sustainable energy systems for remote residential applications ([25], [26], [27]). However, to the best of our knowledge, this report presents the first comparative study that systematically examines these factors, providing a holistic view of alternative energy solutions for data centres.

The following sections outline the research methodology and details of preselecting alternative storage technologies, thoroughly examining their energy performance, environmental impact, and financial assessments. This report aims to equip readers and stakeholders with the knowledge to make informed decisions about adopting alternative energy storage solutions for data centre applications.

2 Methodology

2.1 Methodology for Evaluating Sustainable Storage Technologies

The research followed a sequential exploratory strategy, beginning with a comprehensive literature review on existing studies working on green data centres, followed by a review of different sustainable alternative storage to data centre backups, their global warming potential, levelised costs and total costs of ownership.

Based on an initial literature review, several storage technologies were preselected for further investigation into sustainable data centre backups. This criterion included power rating, discharge time, technological maturity, and the area required for installation. Comparative studies were then conducted on the selected technologies, focusing on crucial energy performance metrics such as efficiency, lifespan, specific energy, and energy density. Additionally, each technology's cradle-to-grave lifecycle assessment was performed to evaluate its global warming potential, offering a comprehensive analysis of the environmental impacts associated with each option.

The Lifecycle Assessment (LCA) quantified the environmental impacts of each technology, focusing on global warming potential. This process involved aggregating and analyzing emissions data throughout the lifecycle of each option. For cost analysis, the Levelized Cost of Energy (LCOE) techniques evaluated the economic costs of different storage solutions, considering initial costs, operations, maintenance, and replacement. The Total Cost of Ownership (TCO) analysis was expanded to include carbon emission costs, providing a comprehensive view of financial impacts by integrating environmental externalities into traditional assessments.

The study rigorously adhered to ethical standards in data collection and analysis, ensuring the confidentiality of any proprietary data and maintaining the integrity of analytical processes. Despite offering significant insights into the sustainability of storage technologies for data centres, there are limitations due to the variability in technology maturity and regional differences in energy sources and costs, which could impact the generalizability of the findings. The methodology facilitated a comprehensive investigation into sustainable storage technologies, effectively balancing environmental, performance, and cost considerations. The results are poised to aid stakeholders in making informed decisions towards adopting greener data centre solutions. In conclusion, this research synthesizes the essential findings and underscores their implications for future research and practical applications in sustainable data centre management.

2.2 Evaluation Methods for Alternative Technologies

2.2.1 Energy Performance: The assessment of energy storage technology performance is critical, particularly when considering the replacement of diesel generators with more sustainable alternatives for backup power in data centres. This evaluation is essential to ensure that any proposed solutions maintain the high levels of reliability and accessibility that diesel generators currently provide. Key performance metrics such as capacity and energy, which reflect the system's ability to store and deliver sufficient power; efficiency and round-trip efficiency, which determine how effectively the storage system conserves energy; response time and ramp rate, which gauge how quickly the system can respond to power demands; and cycle life and degradation, which indicate the longevity and durability of the system, are all carefully analyzed. Additionally, specific energy and energy density are evaluated to ensure the storage solution fits the physical constraints often in data centre environments. Conducting a thorough energy performance assessment helps stakeholders make informed decisions that align with operational requirements and sustainability goals, ensuring that data centres can reliably function without interruption while transitioning away from diesel-based backup systems.

2.2.2 Environmental Impact Assessment: This paper presents a comparative life cycle assessment (LCA) of the global warming potential (GWP) of various storage technologies evaluated cradle-to-grave. Life cycle assessments are influenced by factors such as system description, functional unit selection, impact categories, inventory classification, normalization, and interpretation. We use different units—1 megawatt-hour (1 MWhd) for Diesel Generators and Flywheels and 1 kilowatt-hour (1 kWhd) for other technologies—due to a lack of consistent functional units in the literature for cradle-to-grave studies of GWP. To improve the relevance and comparability of our results against diesel, we scaled the functional units accordingly.

System Description: The ecological loop, shown in Figure 2, illustrates the product life cycle stages. Starting from the "Cradle," where raw materials are harvested and converted into products, the cycle progresses through transportation to the "Gate" (usage stage) and reaches the "Grave" (end-of-life or disposal). At this stage, materials may be recycled and re-enter the loop, promoting sustainability. In each stage, energy and materials enter, while waste and emissions exit, making recycling a crucial factor in reducing ecological impacts.

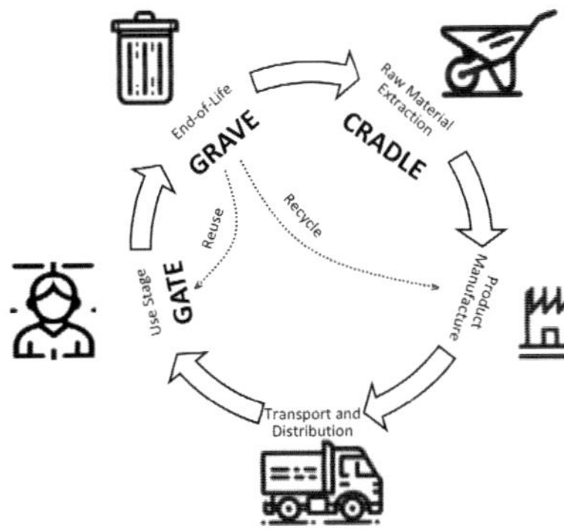

Figure 2: Ecological loop of product life cycle stages

2.2.3 Financial Assessment All the aforementioned technologies are expected to yield equivalent financial returns since they all serve the same purpose of providing backup power, thereby generating similar value. Therefore, the comparative factor among them lies in their respective average costs per kilowatt-hour of electricity produced throughout their operational lifetimes, which essentially reflects the levelised cost of electricity (LCOE). LCOE is widely used in the energy industry to compare the cost-effectiveness of different methods of electricity generation consistently.

$$\text{LCOE} = \frac{\sum_{t=0}^{n} \frac{C_t+M_t+R_t}{(1+r)^t}}{\sum_{t=0}^{n} \frac{E_t}{(1+r)^t}} \tag{1}$$

In this formula:

- C_t represents the capital expenditures in the year t.
- M_t stands for operations and maintenance expenditures in the year t.
- R_t denotes costs associated with component replacements over the system's life.
- E_t is the electricity generation in the year t.
- r is the discount rate.
- n indicates the lifespan of the system.

Total Cost of Ownership
LCOE is a metric used to assess the cost efficiency of an energy-generating system over its operational life. Traditionally, LCOE is calculated without considering the external costs associated with carbon emissions. However, incorporating carbon credit costs provides a more comprehensive understanding of the economic impacts of energy production on the environment.

For a sustainable decision not solely informed by the traditional financial comparison of investments versus returns, TCO_2 stands for Total Cost of Ownership, which includes a carbon price. TU Delft coined the combination of a carbon price and Total cost of ownership (TCO is a financial assessment method that provides for all costs and benefits of the entire lifespan of a product or project.) TCO_2 in its Vision, Ambition and Action Plan for a Sustainable University [28]. The carbon price should cover the impact of (and adaptation and mitigation to) climate change. Current ETS value of a tonne of CO_2-equivalents: ± € 70. Based on various scientific sources, in its sustainable campus programme, TU Delft uses € 150 [29]. Based on a study by the German Ministry of the Environment, to cover all expenses related to carbon compensation, avoidance and recovery of damage done, the price should be close to € 1000/tCO_2eq.

2.3 Preselection of Technologies

Figure 3 classifies energy storage technologies based on power rating, the form of energy stored in, storage duration, and response time ([30], [31]). All technologies in the 50-kW to 10-MW power rating range and high discharge times are considered sustainable alternatives to diesel generators. Figure 4 showcases the technology readiness levels of various alternative storage methods. Only mature technologies are considered for this study. Finally, the area required for storage is considered to finalise the storage technologies. Renewable Diesel - Hydrotreated Vegetable Oil (HVO), Hydrogen energy storage (Electrolyser + Fuel Cell) (HES), Lithium Ion batteries (Li-ion), Sodium Sulphide batteries (NaS), Flywheel-based energy storage system (FBESS), and Compressed Air Energy Storage (CAES) are chosen as alternatives for this study.

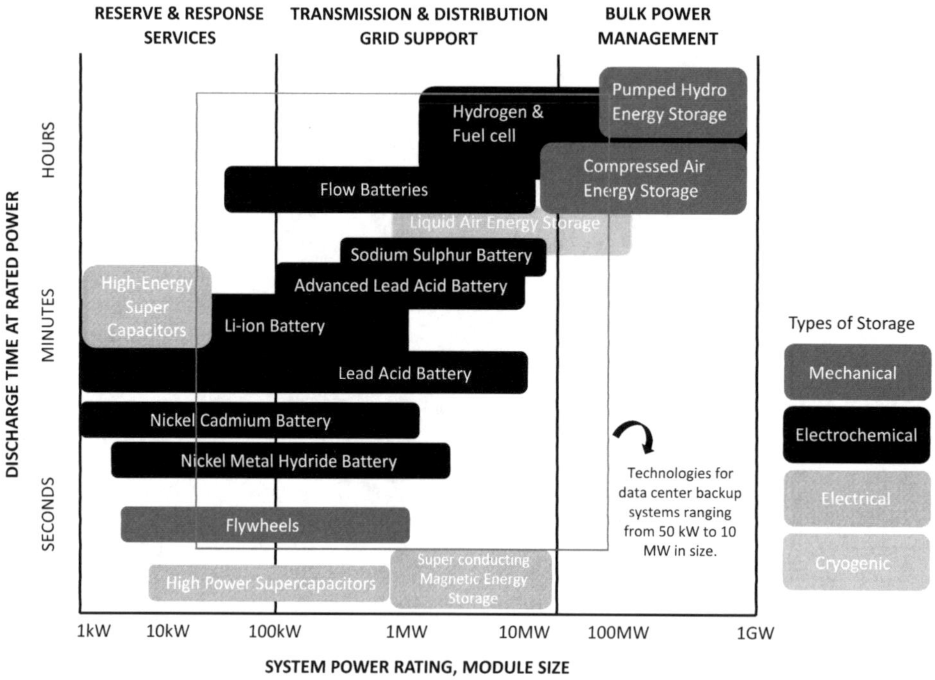

Figure 3: Characteristics of different energy storage technologies adapted from Taylor et al. and Akhil et al. The red square highlights the energy storage systems suitable for backup for data centres. Image adapted from [32]

HVO seamlessly integrates into existing systems but competes with food crops, causing a food-fuel conflict. HES presents a viable option for long-term storage, albeit at a high capital cost and with safety concerns. Both Li-ion and NaS batteries exhibit high energy densities, offering extended discharge times, making them well-suited for backup applications; however, battery manufacturing costs remain prohibitive. Batteries are constrained by charge-discharge cycles, necessitating significant oversizing to address discharge limitations. In contrast, fuel alternatives offer the advantage of refuelling to sustain continuous electricity generation, making them

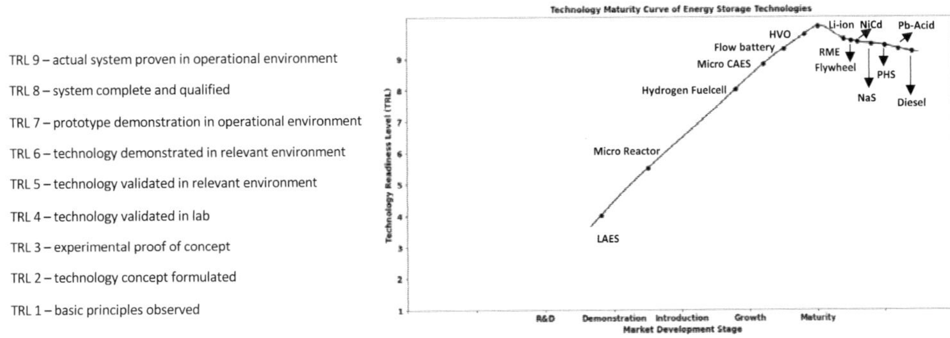

Figure 4: Technology Readiness Levels of energy storage technologies for small scale energy systems. Image adapted from [33]

a reliable power supply source. While CAES holds promise for long-duration storage, it faces geographical limitations in storing compressed air. While CAES systems offer longer-duration storage and potentially lower costs for large-scale applications, their slower ramp-up times compared to batteries could limit certain data centre operations [34]. FBESS, distinguished by its maturity, high efficiency, and energy density, has discharge times ranging from seconds to minutes, limiting its standalone application and favouring hybrid integration with other technologies with slower ramp-up times.

Comparative studies are essential for informed decision-making, resource optimisation, and achieving sustainability goals in data centre operations and other fields. This study offers crucial insights into various storage technologies' energy performance, environmental impact, and financial evaluation. The Energy Performance Analysis assesses storage options based on metrics such as round-trip efficiency, specific energy, discharge time, and lifespan to meet the energy needs of data centres. The Environmental Impact Assessment conducts a comparative Life Cycle Assessment (LCA) of these technologies. Financial analysis includes calculating the Levelized Cost of Energy (LCOE) and lifecycle costs while also considering carbon credit expenses. Despite significant research on sustainable energy solutions for data centres, there is a notable gap in comparative studies focused on alternative storage technologies for backup power. This research addresses this gap by providing a comprehensive comparative analysis of various storage technologies, evaluating their potential as sustainable alternatives to traditional diesel generators for backup power in data centres.

3 Energy Performance Assessment

3.1 Baseline: Diesel Generators

The efficiency values for diesel generators can vary based on the specific model, size, load level, and operating conditions. In general, diesel generators typically have peak efficiencies of 25% to 40% [35]. An industrial diesel generator can last from 20,000 to 40,000 hours of use, equating to approximately 20 to 25 years of operation [36]. Diesel generators boast a quick start-up time of typically 10 seconds or less, ensuring minimal downtime during power outages. In emergency scenarios, these generators can enter operation within a few seconds, providing prompt response to critical events. For routine testing, rapid loading should be avoided, with a recommended gradual loading over 5-15 minutes or more [37].

Attribute	Diesel Generator	Hydrogen Fuel Cell
Peak Efficiency	25% to 40%	45% to 60%
Lifetime (hours of use)	20,000 to 40,000	24000-30000
Lifetime (years of operation)	Approximately 20 to 25	20
Startup Time	Typically 10 seconds or less	10 minutes
Routine Testing Loading	Gradual loading over 5-15 minutes	10 minutes

Table 2: Comparison of Diesel Generator and Hydrogen Fuel Cell Attributes

3.2 Diesel and its alternatives

We compare different fuels, namely diesel and HVO (Hydrotreated Vegetable Oil), both of which can be operated in the same diesel generators. Diesel, a traditional fossil fuel derived from crude oil, which developed over 300 million years from organic remains, exhibits high energy density but emits high levels of pollutants. RME, produced from rapeseed oil, with a much shorter production cycle, offers a cleaner-burning alternative with reduced particulate matter and carbon monoxide emissions, albeit with slightly lower energy density. HVO, synthesized from hydrogenating vegetable oils, shares energy performance characteristics with diesel whilst being a waste product and burning much cleaner. Availability varies, with diesel being widely available but environmentally harmful and renewable diesel being eco-friendly yet dependent on agricultural production and policies. However, it is an emerging market with increasing availability but requires infrastructure development [38] [39].

Table 3: Selected important properties of fuels adopted from [39]

Properties	Lower heating value [$MJ\,kg^{-1}$]	Density at 15 °C [$kg\,m^{-3}$]	Volumetric lower heating value [$MJ\,dm^{-3}$]	Kinematic viscosity at 40 °C [$mm^2\,s^{-1}$]	Cetane number	CFPP [°C]
Diesel	43.0	833.1	36.0	2.50	53	-22
HVO	44.0	781.0	34.0	2.89	95	-32.5
Hydrogen	119.9	0.0838	-	1.21	55	-

3.3 Hydrogen Energy Storage

Hydrogen is a versatile energy carrier with immense potential across various sectors. Its flexibility lies in its ability to be produced from multiple feedstocks through diverse methods, such as electrolysis, steam methane reforming, and biomass gasification. In this study, the focus is on hydrogen production through water electrolysis. There are different electrolysis techniques; Alkaline electrolysed water electrolyser (AEW) & Proton exchange membrane electrolyser (PEM) are the most mature technologies, whose energy performance is listed in table 4.

PEM (Proton Exchange Membrane) fuel cells are highly favoured for backup power applications due to their solid electrolyte, which not only minimizes issues related to corrosion and electrolyte management but also facilitates rapid start-up times. This makes PEM fuel cells exceptionally reliable and efficient for emergency power solutions, ensuring stability and quick response under critical conditions. Essential attributes of PEM fuel cells are mentioned in table 2.

Table 4: Detailed characteristics of AEW, PEM systems adopted from [40]

	AEW	**PEM**
Current density (A cm^{-2})	0.2–0.4	0.6–2.0
Cell voltage (V)	1.8–2.4	1.8–2.2
Cell area (m^2)	<4	<0.3
Operating Temp. (°C)	60–80	50–80
Operating Pressure (bar)	<30	<200
Production Rate^c (m^3 H_2 h^{-1})	<760	<40
Stack energy^c (kWh_{el} m^{-3} H_2)	4.2–5.9	4.2–5.5
System energy^c (kWh_{el} m^{-3} H_2)	4.5–6.6	4.2–6.6
Gas purity (%)	>99.5	99.99
System Response	Seconds	Milliseconds
Cold-start time (min.)	<60	<20
Stack Lifetime (h)	60,000–90,000	20,000–60,000
Lifetime of system (years)	20-30	10-20

Hydrogen offers high specific energy, but it is associated with high flammability, hence has safety concerns regarding storage. Fuel cells are used to convert Hydrogen to electricity, and Hydrogen showcases efficient conversion of chemical to electrical energy. Enhanced storage solutions and robust fuel cell technologies improve its longevity and reliability. Low self-discharge rates ensure prolonged availability, which is crucial for long-term storage applications. In addition, the rapid response times of fuel cells make Hydrogen suitable for quick start-up [41].

3.4 Batteries

3.4.1 Lithium Ion Lithium, renowned for its remarkable reactivity and energy storage capabilities, is pivotal in advancing Lithium-ion (Li-ion) battery technology. Its unique properties, including low atomic weight and small atomic radius, empower Li-ion batteries with high voltage and charge storage capacities per unit mass and volume, enabling rapid and efficient energy transfer and storage. Various electrode materials, such as lithium cobalt oxide, graphite, lithium manganese oxide, and lithium iron phosphate, cater to diverse applications, from portable electronics to electric vehicles [42, 43].

Li-ion batteries boast one of the highest energy densities among commercial battery technologies, making them versatile for various applications, especially power-intensive ones like transportation. With high efficiency and extended lifetimes of up to 15 years, Li-ion batteries offer reliability and longevity. They exhibit low self-discharge rates, minimal maintenance requirements, and no memory effects, ensuring sustained performance. However, challenges such as supply chain stress due to rare materials, high costs, and scalability issues in storage systems present notable hurdles for widespread adoption [42].

3.4.2 Sodium Sulphur NaS batteries are touted for their long-duration energy storage. Their superior performance stems from extensive research and commercial use over two decades. They boast high energy density, efficiency, and long life, and they are made from cost-effective and environmentally friendly materials. A notable feature of these batteries is zero daily self-discharge rates [44]. However, NaS batteries increase operational costs due to the requirement of higher operating temperatures (around 300-350°C to liquefy Na). They are prone to safety concerns due to the reactivity of sodium and sulfur. Therefore, they are best suited for stationary applications where safety precautions can be implemented effectively. Consequently, they function well as high-capacity grid solutions, offering reliable long-term energy storage, and can serve as alternatives to fossil fuel-based power generators [45]. The battery characteristics are given in Table 5.

Table 5: Battery Characteristics Comparison

Characteristics	Li-ion	NaS
Cell voltage	2.5 – 5 V	1.8 – 2.71 V
Specific energy	80 – 250 Wh/kg	150 – 240 Wh/kg
Energy density	95 – 500 kWh/m^3	150 – 350 kWh/m^3
Efficiency	75 – 97 %	75 – 90 %
Working temperature	20 – 65 °C	300 – 350 °C
Lifetime cycles	100 – 10000	2500 – 40000
Lifetime	5 – 15 years	10 – 15 years
Max. depth of discharge	100 %	100 %
Self-discharge rate	0.1 – 0.3 % per day	0 % per day
Power rating	0 – 0.1 MW	0.05 – 34 MW

3.5 Compressed Air Energy Storage
CAES systems exhibit notably low energy and power densities owing to factors such as closed-cycle operation, logarithmic pressure variations, energy losses during compression/expansion, pressure fluctuations, system configuration, and heat transfer inefficiencies necessitating extensive storage facilities for effective operation, such as underground salt caverns such as those in McIntosh and Huntorf. This makes CAES primarily suitable for large-scale, stationary energy storage applications. Additionally, CAES demonstrates impressive capabilities in storing and delivering substantial energy and power with minimal self-discharge over extended periods. Moreover, CAES facilities boast long lifespans and numerous charge-discharge cycles, offering cost-effective solutions despite high initial capital investments. However, CAES systems currently face challenges regarding roundtrip efficiency, typically ranging from 42% to 89%. While advancements like adiabatic and isothermal CAES hold promise for enhancing efficiency, further improvements are needed to rival Pumped Hydro Storage (PHS) plants in scale and efficiency while reducing environmental impact [46], [47], [48].

Characteristic	Unit	Value (range)
Energy density	Wh/L	2–6
Specific energy	Wh/kg	30–60
Power rating	MW	5–400
Rated energy capacity	MWh	580–2860
Daily self-discharge	N/A	Small
Lifetime	Years	20–60
Cycling times	Cycles	8000–30,000
Cycle efficiency	%	42–89
Storage duration	N/A	Hours — months
Discharge time	Hours	1–24+

Table 6: Energy Storage System Characteristics adopted from [48]

3.6 Flywheel-based energy storage systems

FBESS are known for their remarkable energy efficiency, which is attributed to advances in material science, minimal friction losses, and negligible wind resistance. Compared to batteries, FBESS demonstrate exceptional longevity and can endure numerous charge/discharge cycles without being affected by temperature or depth of discharge (DOD). Unlike batteries, FBESS are operationally reliable as their mechanical nature leads to fewer failure modes; disposal raises no environmental concerns as they do not contain any harmful chemicals. Monitoring the state of charge (SOC) for FBESS is straightforward, relying solely on measuring flywheel spinning speed. High-speed rotating masses require precision engineering and advanced materials, which escalate the complexity and cost of FBESS. The energy-to-weight ratio of FBESS tends to be less favourable than alternative storage options, limiting their application in mobile applications, and flywheels have high self-discharge rates, making them suboptimal for long-term energy storage [49], [50]. The FBESS characteristics are given in Table 7

Table 7: Flywheel Energy Storage Characteristics [50], [49]

Feature	Value
Efficiency (%)	85–95
Specific energy (Wh/kg)	5–150
Power rating (MW)	0.1–4200
Charging time	minutes
Discharge time	15 s–15 min
Discharge depth	deep
Response time	milliseconds
Service life (Year)	$\geq$ 20
Maintenance cycles	$\geq$ 10 years
Operating temperature (°C)	-40–50
Daily self-discharge	5%

Figure 5 (a) provides a comparative bar chart detailing the specific energy and energy density of Diesel, Renewable Diesel, and Hydrogen. Figure 5 (b) visually represents the energy performance of various storage technologies, illustrating key attributes such as efficiency, response time, self-discharge rate, and lifespan across technologies like Diesel Generators, Hydrogen Fuel Cells, CAES, and others.

4 Environmental Impact Assessment

4.1 Diesel Generators

4.1.1 Diesel Eric A. Alsema [51] conducted a Life Cycle Assessment (LCA) on diesel generators, focussing on generating 1 MWh of electrical energy. The study aimed to evaluate diesel generators' environmental impacts for domestic electrification. The scope included analysing a 5 kVA diesel generator with a 25% fuel-to-electricity conversion efficiency, considering fuel transportation over 100 km and a 10-year life expectancy. Inventory analysis quantified material use, energy consumption, and emissions, while impact assessment utilized the Eco-indicator '95 method to assess Global Warming Potential [52].

The analysis highlighted in Table 8 shows that the primary environmental impact arises from emissions generated during fuel combustion, with a secondary influence attributed to the processes involved in fuel extraction and refining. Similar fuel combustion values are supported by a study by Volker Quaschning and Bernhard Siegel [53]. In contrast, the environmental ramifications associated with the manufacturing of the generator itself and the transportation of fuel over

Journal of Physics: Conference Series **2929** (2024) 012008 doi:10.1088/1742-6596/2929/1/012008

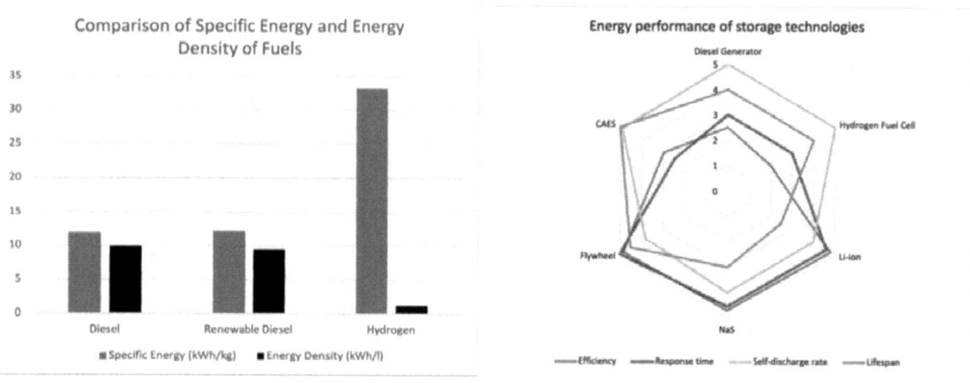

(a) (b)

Figure 5: (a) Specific Energy and Energy Density of Fuels (b) Energy Performance of Storage Technologies (adopted from tables 3,4, 5, 6, 7)

Class	Unit	Total	Diesel Genset	Fuel Production	Fuel Transport	Fuel Combustion
GWP	kg CO_2	1.27E+03	4.4	205	4.62	1.05E+03

Table 8: LCA results of diesel generator adopted from [51]

a distance exceeding 100 kilometres were found to be relatively minimal. The emission factor applied to the diesel generator amounted to 1.27 x 10^3 kilograms of CO_2 per megawatt-hour (kg CO_2/MWh). A change in life expectancy and distance wouldn't impact the emissions. However, fuel combustion is inversely proportional to efficiency, and an efficiency of 40% could reduce the emissions to approximately 0.67 kg CO_2 and fuel production to 0.128 kg CO_2.

The above analysis offers a cradle-to-gate Life Cycle Assessment (LCA) of a diesel generator set. Thus, Benton et al.'s [54] investigation into the life cycle energy assessment of a standby diesel generator set has been consulted to address the disposal phase of a diesel generator. Scenario 2 (Landfill 34%, Recycle 34%, Remanufacture 32%) is chosen for analysis due to the uncertainty of the disposal route. As a midpoint between zero and complete recycling, Scenario 2 provides a balanced approach suitable for cradle-to-grave analysis, making it the most relevant scenario for our study's focus. **Scenario 2** results in a **-52.20%** change in materials energy, i.e. by remanufacturing, the product can lead to a 52% reduction in the initial energy consumption.

4.1.2 Renewable Diesel The studies conducted by Miguel Brandão et al., [55] [56], have been utilised to evaluate the environmental impacts associated with biofuel production. The first paper [56] discusses the modelling approach, while the second paper [55] quantifies the carbon footprint of biofuels using the openly accessible biograce tool. The study utilises four modelling approaches to evaluate biofuel carbon footprints, including the Renewable Energy Directive (RED), Attributional LCA (ALCA), and Consequential LCA (CLCA). These methods assess direct and indirect land-use changes, environmental impacts, and system-wide effects of biofuel production. The biofuel pathway involves multiple stages, including cultivation, processing, and combustion of biofuels as an alternative to fossil fuels.

Several vital stages are involved in the production process of renewable diesel from rapeseed, beginning with the cultivation activities, such as the growth and harvesting of rapeseed, followed by the drying of the harvested crop. Subsequently, during the processing phase, the oil undergoes essential procedures, including oil extraction from rapeseed and the subsequent hydrogenation of vegetable oil. Transportation plays a pivotal role in the logistical chain, encompassing the movement of rapeseed from cultivation sites to processing facilities, transportation of rapeseed oil, and the delivery of hydrogenated vegetable oil (HVO) to depots. Finally, transporting the final product to filling stations completes the chain. Each stage contributes to the overall environmental impact and sustainability of renewable diesel production from rapeseed. Our analysis assumes that any carbon dioxide captured during feedstock cultivation is offset by the emissions released during combustion. Consequently, we consider the carbon emissions during the usage phase negligible or zero.

4.2 Other technologies

Oliveira et al. [57] evaluated the environmental impacts of integrating storage systems into the electricity grid, comparing various energy storage technologies on a standardised functional unit: one kilowatt-hour of stored and delivered electricity utilising existing studies and databases for inventory data. It encompasses the entire lifecycle, cradle to grave (from raw material extraction worldwide to assembly and use in Belgium, including end-of-life disposal). Distribution network effects are excluded, focussing solely on system outputs. Energy mixes utilised include Belgium 2011 (Belgium electricity mix for 2011), UCTE 2004, 100% wind, and 100% photovoltaic. The energy mixes are shown in table 9. Impact assessment employs ReCiPe 2008 methodology, analysing relevant midpoint categories and providing a single-score evaluation via SimaPro 7.3.3, Commercial Analyst version.

Table 9: Energy Mixes

Production unit	**Nuclear**	**Natural gas**	**Coal**	**Municipal waste**	**Blast furnace gas**	**Wind**	**Hydro**	**Biomass**	**Oil**
BE2011	59.0	28.0	4.60	2.60	2.30	1.10	1.70	0.69	0.00
UCTE 2004	17.00	21.00	30.00	1.0	1.00	7.00	15.00	1.00	7.00

4.2.1 Hydrogen: The scope of hydrogen-based power generation encompasses various production methods, with electrochemical processes being prominent, employing electrolyzers to split water into Hydrogen and oxygen. In this study, the focus lies on proton exchange membrane fuel cells (PEMFC), a mature technology widely applied in commercial settings. These fuel cells utilize high-pressure Hydrogen derived from high-temperature electrolysis as a feedstock. The inventory for the fuel cell stack and electrolyzer is drawn from the NEEDS European Project database. Components crucial to hydrogen production include electrolyzers, diaphragm compressors, storage modules, and structural elements like walls and foundations, along with considerations for use and maintenance. For fuel cells, key components and maintenance aspects encompass the stack, balance of plant (BOP), potential reformers for methane feedstock, and requisite support structures. This integrated approach underscores the comprehensive evaluation and management of hydrogen-based power generation systems.

4.2.2 Li-ion & Sodium Sulphur batteries: The life cycle inventories used to model the battery systems are sourced from the SUBAT project on sustainable batteries [58] for Li-ion batteries and from [59] for the sodium-sulfur batteries. Li-ion batteries were found to have significant impacts during the manufacturing stage, attributed to mining activities for copper and lithium and energy requirements for production. The charge/discharge efficiency directly influenced the impact of the use stage. NaS batteries showed similar patterns, with manufacturing impacts from mining activities and energy needs. Renewable energy mixes generally performed better than average electricity mixes, except for scenarios where the manufacturing burdens of solar panels offset gains from energy storage. Li-ion and NaS batteries performed below the Belgian mix threshold when storing wind energy.

4.2.3 CAES: The emissions analysis for Compressed Air Energy Storage (CAES) systems reveals distinct contributors to greenhouse gas (GHG) emissions. In CAES, natural gas combustion in the turbine constitutes the primary source of GHG emissions, while construction-related activities contribute less than 2% of total emissions. Turbomachinery and air storage mediums contribute significantly to the total life cycle of GHG emissions for CAES infrastructure. Notably, the environmental impact of electricity consumption during the use phase is substantial, emphasising the importance of transitioning to cleaner energy sources. Efforts to optimise construction processes and reduce material requirements could further mitigate the environmental footprint of CAES technologies. Table 1 summarises the emissions range for CAES and systems.

Table 10: Total CO_2 Emissions for Li-ion and NaS Batteries Across Energy Mixes

Energy Mix	**UCTE 2004**	**Belgium 2011**	**PV**	**Wind**
HES (kg CO_2-eq./kWh)	1.62	0.58	0.29	0.05
Li-ion (kg CO_2-eq./kWh)	0.6	0.25	0.15	0.068
NaS (kg CO_2-eq./kWh)	0.65	0.24	0.13	0.034
CAES (kg CO_2-eq./kWh)	0.75	0.27	0.14	0.024

4.3 Flywheel

The study by Md Mustafizur Rahman [60] compares steel and composite rotor flywheel energy storage systems, assessing their environmental performance through net energy ratio (NER) and greenhouse gas emissions with a functional unit as 1 MWh of electricity delivered over a 20-year project lifetime. The assessment includes inventory analysis of energy/material inputs/outputs and impact assessment on environmental performance.

The life cycle assessment (LCA) of flywheel-based energy storage systems examines various phases contributing differently to greenhouse gas (GHG) emissions and energy consumption, including material production, manufacturing, operation, transportation, and end-of-life (EOL). Notably, the operation phase significantly contributes to GHG emissions, accounting for 60-75% for steel rotor FESS and 54-76% for composite rotor FBESS, with standby mode emissions contributing 34-70% and 17-48%, respectively. Material production emissions are 7.7 kg-CO_2eq/MWh for steel rotor FESS and 17.9 kg-CO_2eq/MWh for composite rotor FESS. In comparison, manufacturing emissions are 19.1 kg-CO_2eq/MWh and 1.7 kg-CO_2eq/MWh, respectively, with steel rotor manufacturing being more GHG-intensive. End-of-life emissions contribute 3-6% to the life cycle GHG emissions. Additionally, Table 11 compares flywheels' life cycle environmental impact with different electricity sources for charging, indicating that composite flywheels are a more environmentally friendly option for energy storage regarding carbon emissions. This assessment highlights the importance of considering various life cycle phases in evaluating the environmental performance of flywheel energy storage systems [61, 60].

The life cycle assessment values for global warming potential of various energy technologies, as shown in Figure 6(a), clearly demonstrate the environmental advantages of using wind energy,

Source of Electricity for charging	Solar	Wind	Grid	Mixed
Grid emission factor (kg-CO_2eq/MWh)	26 - 183	3 - 45	590	
Steel Flywheel (kg-CO_2eq/MWh)	121.4	95	649	288.5
Composite Flywheel (kg-CO_2eq/MWh)	75.2	48.9	623	249

Table 11: Life cycle carbon footprint of Flywheel [60, 61]

which results in the lowest CO_2 emissions. In stark contrast, energy systems dominated by fossil fuels contribute significantly higher emissions. This difference highlights the urgent need for a shift toward renewable energy sources. Further insights from Figure 6(b) show that the use phase—which is influenced by system efficiency and the energy mix—significantly impacts total emissions. This finding strengthens the case for adopting sustainable storage solutions, especially when paired with green energy generation. Such a move not only reduces environmental impact but also supports a broader shift towards sustainable energy practices.

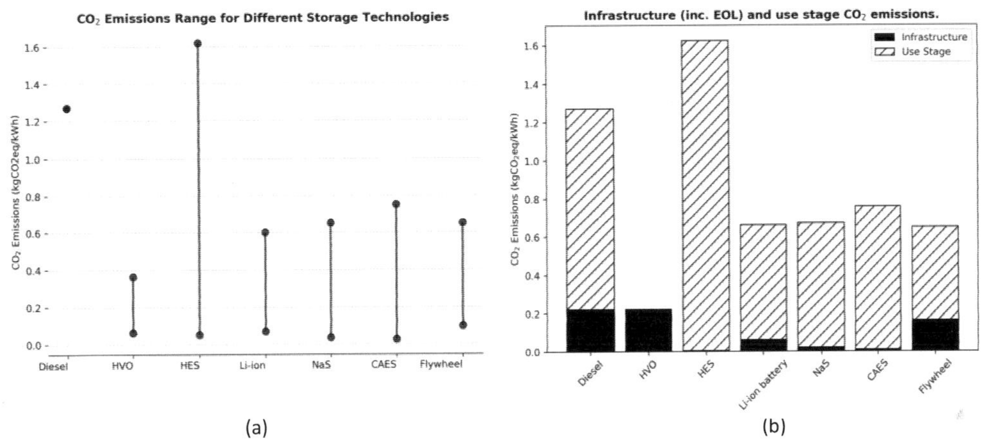

Figure 6: (a) GWP of various storage technologies (b) Infrastructure and Use phase CO_2 emissions

5 Financial assessment

5.1 Capital Costs:

The upfront capital expenditure denoted as $C_{\text{cap}}^{\text{initial}}$ (€), for an energy storage (ES) system encompasses all costs associated with its core components. These include storage containers, power converters, transformers, protective mechanisms, cooling systems, and other related hardware. This initial capital investment is categorically divided into three primary segments, as shown in equation 2:

- C_{SC} represents the cost of the storage container.
- C_{PC} covers the expenses for the power conversion systems.
- C_{BP}, or the balance of plant costs, pertains to the expenditures for protective devices, cooling systems, and other ancillary components.

Journal of Physics: Conference Series **2929** (2024) 012008 doi:10.1088/1742-6596/2929/1/012008

The equation representing these contributions is:

$$C_{\text{cap}}^{\text{initial}} = C_{\text{SC}} + C_{\text{PC}} + C_{\text{BP}} \tag{2}$$

The annualized capital cost relative to the power capacity of the ES system (P), referred to as C_{t} (€/kW-year), provides a normalized metric to assess the capital efficiency per unit of power capacity, facilitating comparative analyses. The capital cost itself (C_t) can be detailed further:

$$C_t = \frac{C_{\text{cap}}^{\text{initial}}}{P} \times RF \tag{3}$$

Where RF is the recovery factor, calculated as:

$$RF = \frac{(1+r)^n - 1}{r \times (1+r)^n} \tag{4}$$

Here, r represents the discount rate and n the project's lifespan.

5.2 Operational and Maintenance Cost Estimations

Fixed and variable costs during the operational phase are crucial in overall cost calculations. The annual fixed costs do not depend on the operation phase and are typically lower, whereas variable costs can fluctuate based on the usage and maintenance requirements of the system. The total operational and maintenance cost, inclusive of fuel and other variable components when applicable, can thus be summarized:

$$M_t = C_{FOM} + C_{VOM} \tag{5}$$

Where C_{FOM} and C_{VOM} represent the fixed and variable operational and maintenance costs, respectively.

The annualized fixed cost per power capacity (€/kW-year) of an energy storage (ES) system can be computed by employing the annualizing factor (F_A), which incorporates the variation rate of fixed operation and maintenance costs (V_{FC}). The formula for the annualizing factor, which takes into account these operational costs, is provided below:

$$F_A = RF \sum_{m=1}^{n} \left(\frac{1 + V_{FC}^m}{1 + r^m} \right) \tag{6}$$

This equation reflects the compounded impact of the fixed operation and maintenance costs over the lifetime of the ES system, and m spans from 1 to n years of operation.
The fixed operational cost of an energy storage (ES) system, represented as C_{FOM}, integrates the annualized fixed costs and includes considerations for short-term operational scenarios. The primary calculation is outlined below:

$$C_{\text{FOM}} = C_{\text{fixed}} \times F_A + C_{\text{FOM}}^{\text{ST}} \tag{7}$$

where C_{fixed} is the fixed operation and maintenance cost coefficient.
For ES technologies intended for short-term applications, additional costs must be considered to cover the energy required to maintain readiness and compensate for self-discharge losses:

$$C_{\text{FOM}}^{\text{ST}} = B_{\text{elec}} \left(\frac{d}{24} \right) \left(\frac{S_{\text{dis}}}{100} \right) \times F_B \tag{8}$$

$$F_B = RF \sum_{m=1}^{n} \left(\frac{(1 + V_{\text{elec}})^m}{(1+r)^m} \right) \tag{9}$$

Journal of Physics: Conference Series **2929** (2024) 012008 doi:10.1088/1742-6596/2929/1/012008

where B_{elec} is the market energy price, S_{dis} is the self-discharge rate, d is the number of working days, and V_{elec} is the rate of variation in energy prices.
Additionally, the variable operation and maintenance costs are formulated to reflect hourly operations:

$$C_{\text{VOM}} = B_{\text{elec}} \left(\frac{d}{24}\right) \left(\frac{h_d}{\mu}\right) \times F_B \tag{10}$$

For technologies that use natural gas, such as compressed air energy storage, the cost calculation is extended to include gas costs:

$$C_{\text{VOM}} = \left[B_{\text{elec}} \left(\frac{d}{24}\right) \left(\frac{h_d}{\mu}\right) \times F_B\right] + \left[B_{\text{gas}} \left(d\right) \left(h_d\right) \times G_r \times \frac{F_{\text{gas}}}{10^3}\right] \tag{11}$$

$$F_{\text{gas}} = RF \sum_{m=1}^{n} \left(\frac{(1+V_{\text{gas}})^m}{(1+r)^m}\right) \tag{12}$$

where h_d represents the operating hours per day, B_{gas} and G_r are the per unit gas cost and gas price factor, respectively, and V_{gas} represents the rate of variation in gas prices.

5.3 Replacement Costs

An energy storage project's service life typically exceeds its storage containers' operational lifespan due to ageing, technological wear and tear, and usage patterns. Consequently, the replacement costs for the storage containers are a critical component of the overall cost model for ES technologies. However, it is essential to note that replacement costs do not always extend to other system components, such as converters or balance of plant equipment, which may last for the entire duration of the project. The formula for calculating the annualized replacement costs is given as follows:

$$R_{\text{t}} = B_{\text{rep}} \times \frac{h \times N}{\mu} \times F_{\text{rep}} \tag{13}$$

where R_{t} represents the total replacement costs, B_{rep} is the unit replacement cost, h is the number of hours the component is operational per replacement cycle, N is the number of replacements during the project's life, and F_{rep} is the annual replacement cost factor.
The annual replacement cost factor, F_{rep}, is calculated as:

$$F_{\text{rep}} = RF \sum_{m=1}^{N} \left(\frac{1}{(1+r)^{m \times A}}\right) \tag{14}$$

where N is the number of replacements, and A is the period until the replacement of the ES is required.

5.4 Formulation of the cost model

In this work, the cost model for different types of energy storage (ES) technologies, which includes detailed economic metrics and operational factors, is developed in Python to ensure accurate calculations, effective data management, and rigorous testing, all of which guarantee the proper functioning of the model. Figure 7 shows a schematic diagram of the proposed cost model of the ES technologies.

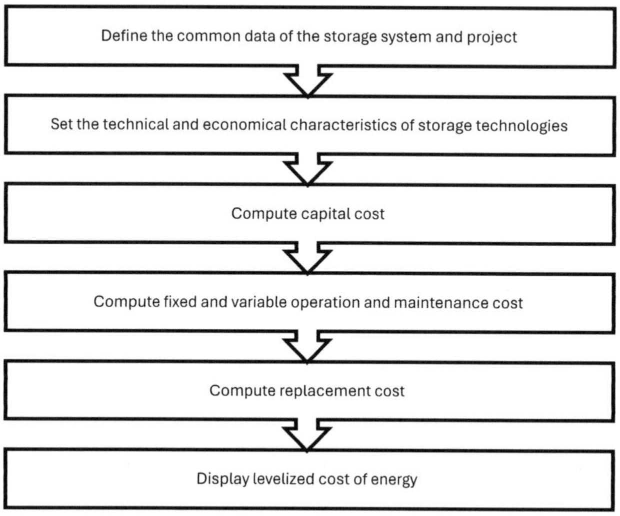

Figure 7: Schematic diagram of the proposed cost model of the ES technologies.

Table 12 provides the typical data and coefficients along a project lifetime, such as the discount rate, electricity price, rate of variation of the electricity cost, and rate of variation of the fixed operation and maintenance cost. The annual average electricity prices for the Netherlands for the year 2023 have been used for this study, which turns out to be 0.096 (€/kWh), which is rounded to 0.1 [62]. Similarly, the prices for diesel and renewable diesel have been taken for 2023 [63]. Commercially available values are considered for diesel-generator capital and maintenance costs. The storage systems have been sized to provide 24 hours of backup for a 1MW system.

Table 12: Common Data and Coefficients

Parameter	Value
Lifespan of the project (years)	25
Market Price of Electricity (€/kWh)	0.1
Market Price of Diesel (€/kWh)	0.6
Market Price of HVO (€/kWh)	0.75
Discount rate	0.085
Rate of variation of Belec	0.03
Rate of variation of gas price	0.03
Rate of variation of FOM cost	0.01

Based on [64], [65], [66], [67], [68], [69], [70] the technical parameters required to perform the cost model are summarised in Table 13

Table 13: Energy Storage Configurations

Parameter	HES	Li-ion	NaS	CAES
P (Power Capacity MW)	1000	6000	4000	1000
E (Energy Capacity MWh)	25000	25000	25000	25000
c_s (Cost Storage €/kWh)	3.7	795	298	40
c_{pc} (Cost Power Conversion €/kW)	2465	463	366	843
c_{bp} (Cost Balance Plant €/kW)	25	80	80	15
μ (Efficiency %)	45	90	90	75
$DOD_{\max}$ (Max Depth of Discharge %)	100	80	80	60
c_{fixed} (Fixed Cost €/kW)	25	6.9	3.6	3.9
n (Cycle Life)	20000	4500	4500	250000
A (Age in Years)	15	10	12	40
N (Number of Replacements)	1	2	2	0
R_t (Replacement Cost)	413	795	298	0
d (Operating Days per Year)	365	365	365	365
S_{dis} (Self Discharge Rate)	0	0.002	0.2	0
G_r (Gas Rate Factor MJ/kWh)	-	-	-	4.43
B_{gas} (Gas Price €/GJ)	-	-	-	9.97

Diesel generators, traditionally employed for backup purposes, are characterized by their notably infrequent use, typically limited to approximately 14 hours annually (0.0384 hours daily average), primarily for testing and maintenance checks. Figure 8 delves into the LCOE across various technologies, examining the economic implications of different average daily operational hours.

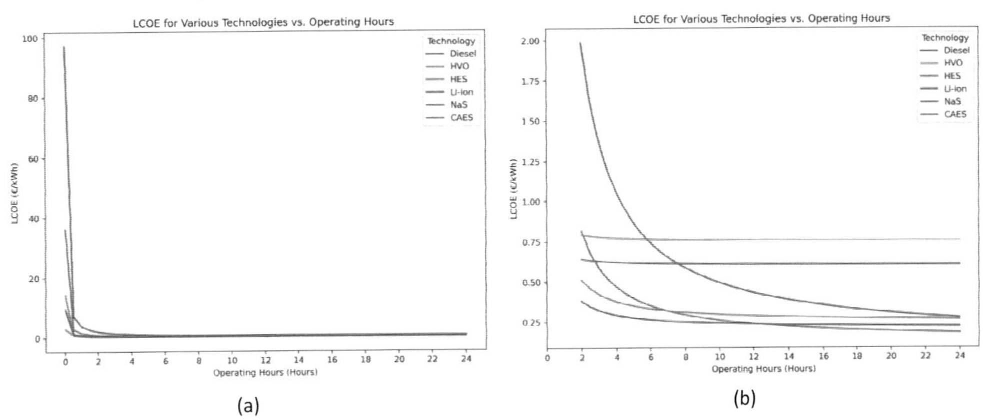

Figure 8: LCOE values of the ES technologies for operating hours in the range (a) 0 to 24 hours (b) 2 to 24 hours

From Figure 8(a), it is observed that the LCOE significantly decreases with even modest increases in the average daily operating hours. As shown in Table 13, the sizing requirements for batteries in terms of megawatts (MW) are higher than other technologies. This discrepancy arises because batteries possess inherently lower energy capacities due to their charge-discharge characteristics. In contrast, other technologies benefit from the ability to store fuel and utilize it as needed, offering a more flexible energy management approach. This leads to higher LCOE values for batteries. Figure 8(b) illustrates the LCOE across various technologies with daily average operating hours commencing from 2 hours. As previously noted, while batteries initially exhibit a higher LCOE, their costs decrease relative to diesel generators as operating hours increase (NaS has the least LCOE for a higher daily average operating hours). This cost efficiency is attributable to the higher efficiency of batteries and the economic advantage of electrical energy, which is significantly cheaper than diesel and HVO. A further decrease in the price of batteries will make a strong case for them. Notably, CAES is the most cost-effective technology across the observed range, followed by HES.

5.5 TCO_2

The social cost of carbon is incorporated, providing a more comprehensive understanding of the economic and environmental impacts of energy production. This adjusted calculation is called the TCO_2, which sums up the traditional LCOE and the costs associated with purchasing carbon credits. The formula for the TCO_2 is:

$$\mathrm{TCO_2} = \mathrm{LCOE} + C_{\mathrm{carbon}} \tag{15}$$

Table 14: LCOE and CO_2 Emissions for Various Energy Technologies

Technology	LCOE (€/kWh)	CO_2 Emissions Min (kg/kWh)	CO_2 Emissions Max (kg/kWh)
Diesel	0.64	1.27	1.27
Renewable Diesel (HVO)	0.79	0.061	0.363
Li-ion	1.98	0.068	0.6
NaS	0.81	0.038	0.65
HES	0.51	0.034	1.62
CAES	0.38	0.024	0.74

Table 14 presents the LCOE values for various storage technologies, assuming an average daily operation of two hours and the life cycle CO_2 emissions range. The minimum values indicate scenarios utilizing renewable energy sources, whereas the maximum values are derived from a predominantly fossil fuel-based energy mix.

The carbon credit costs required for various technologies to achieve the exact levelized cost of energy (LCOE) as diesel generators are specified as follows: For Renewable Diesel (HVO), the carbon price needs to range from €124.07/tonne to €165.38/tonne to match diesel's LCOE. Li-ion technology requires a significantly higher carbon price, ranging from €1,114.81/tonne to €2,000.00/tonne, to align with diesel's LCOE. For Sodium-sulfur (NaS) batteries, the carbon price varies between €137.99/tonne and €274.19/tonne to achieve equivalence. Hydrogen Energy Storage (HES) and Compressed Air Energy Storage (CAES) systems already exhibit lower LCOEs than diesel and have lower emissions, resulting in a consistently lower total cost of ownership than diesel generators.

Particularly for HES, if derived from green Hydrogen (produced using renewable energy sources), emissions are significantly lower than diesel generators. However, when the majority of energy used in hydrogen production comes from fossil fuels, emissions can be relatively high due to the low round-trip efficiency of the process. A carbon credit cost of €371.00/tonne is necessary for HES to equate its total carbon cost to diesel generators. Regarding HVO, current carbon prices already present a competitive alternative to diesel. As the production of HVO scales up and costs potentially decrease, it could become even more economically attractive, enhancing its market competitiveness.

Diesel generators, HES, and CAES systems typically require about 15 minutes for start-up. During this period, batteries have been traditionally employed to bridge the gap. However, FBESS emerge as a promising alternative due to their high efficiency and rapid response capabilities. For the initial start-up phase, the Levelized Cost of Energy (LCOE) for Sodium-sulfur (NaS) batteries, FBESS, and Lithium-ion (Li-ion) batteries increases in the mentioned order. It is important to note that the CO_2 emissions generated during this brief start-up phase are minimal and do not significantly affect the LCOE calculations.

6 Conclusion and Discussions

6.1 Key Findings

Renewable diesel has specific energy and energy density comparable to traditional diesel, allowing its use in existing diesel generators without modifications. This compatibility eases the transition to sustainable fuel options within established infrastructures. Hydrogen boasts a high specific energy but has a lower energy density. Its production versatility and adaptability across applications are significant advantages. However, hydrogen energy systems typically exhibit lower efficiency, requiring technological improvements for greater viability. Batteries are known for high efficiency and rapid response times, essential for applications needing quick energy dispatch. Their shorter lifespan compared to other storage technologies may limit long-term applications. Both batteries and flywheels provide immediate power during outages, but their energy storage capacity must be carefully sized to meet demand during critical backup periods. Their operation involves cyclic charging and discharging, which restricts available energy capacity at any moment. In contrast to fuel-based technologies, the operational capacity of batteries and flywheels is determined by their maximum stored energy and discharge rates. Flywheels, with low discharge times, can effectively replace batteries for the initial start-up of diesel generators, providing a quick energy-release solution. Compressed Air Energy Storage (CAES) has a long lifespan and is suitable for medium-term storage, but its feasibility is contingent on specific geographic conditions due to storage location requirements. CO_2 emissions primarily occur during the use phase of these technologies. The efficiency of diesel generators directly affects emissions, with higher efficiency resulting in lower emissions. For other technologies, emissions depend significantly on the energy mix; systems powered by renewable sources emit far fewer emissions than those relying on fossil fuels. The Levelized Cost of Energy (LCOE) for storage systems varies with daily operational hours. CAES is cost-effective across different operating scenarios, while batteries and hydrogen energy systems (HES) face higher initial costs due to complex infrastructure needs. Nonetheless, HES has a lower LCOE than batteries, attributable to the higher sizing requirements of batteries. Fuel costs for renewable diesel significantly influence its LCOE. Overall, renewable diesel and CAES are scalable and cost-effective. The high upfront costs of batteries and HES underline the necessity for further research and development to reduce costs and enhance efficiency and lifespan.

6.2 Conclusions

In data centres that currently rely on diesel generator backups, transitioning to renewable diesel (HVO) presents a viable option due to its seamless integration with existing systems. While renewable diesel may be costlier initially, it is less carbon-intensive and could become more economical as production increases. Biodiesels are, therefore, crucial for the energy transition. Compressed Air Energy Storage (CAES) stands out for its lower lifecycle emissions and operational costs, making it an excellent choice when geographical conditions are favourable; if not, Hydrogen Energy Storage (HES) serves as a strong alternative with nearly comparable benefits.

Although batteries have a higher Levelized Cost of Energy (LCOE), their economic viability improves with extended daily operational hours, allowing them to potentially match or exceed the cost-effectiveness of diesel generators. However, extending backup times to three days significantly increases both capital and operational costs for batteries—tripling them compared to other technologies, which primarily face only rising fuel costs. Additionally, longer backup times necessitate more space to accommodate the larger battery systems. This emphasizes the importance of backup duration in technology selection for data centres. Both batteries and flywheels have unique characteristics and cost implications, yet their integration can substantially enhance operational resilience, ensuring reliable power during the start-up of conventional diesel generators.

6.3 Recommendations

Existing data centres using diesel generator backups should transition to renewable diesel to reduce environmental impacts. For new data centres, it is crucial to implement on-site renewable energy production or procure green energy to minimize CO_2 emissions. The choice of renewable energy should consider geographic and climatic conditions; for example, wind energy is effective in temperate and coastal regions, while solar energy performs best in hot, dry climates. Economically, designing storage systems to operate for an average of two hours daily may necessitate oversizing capacities. Emerging technologies like Liquid Air Energy Storage (LAES) and Small Modular Reactors (SMR) offer scalable and efficient backup solutions for hyper-scale data centres. These strategic implementations not only improve energy efficiency but also contribute to global efforts to reduce carbon footprints, guiding data centres towards more sustainable operations.

Bibliography

[1] Beth Whitehead et al. "Assessing the environmental impact of data centres part 1: Background, energy use and metrics". In: *Building and Environment* 82 (2014), pp. 151–159.

[2] Meta. *Meta Datacentre.* Accessed on 12 Feb 2024. URL: `https://datacenters.atmeta.com/`.

[3] Adobe stock. *Server room interior in data center.* Accessed on 12 Feb 2024. URL: `https://stock.adobe.com/nl/search?k=datacenter`.

[4] Donald L Beaty. "Internal IT load profile variability". In: *ASHRAE Journal* 55.2 (2013), pp. 72–75.

[5] Eduard Oró et al. "Energy efficiency and renewable energy integration in data centres. Strategies and modelling review". In: *Renewable and Sustainable Energy Reviews* 42 (2015), pp. 429–445.

[6] Eric Masanet et al. "Recalibrating global data center energy-use estimates". In: *Science* 367.6481 (2020), pp. 984–986.

[7] Hannah Ritchie, Pablo Rosado, and Max Roser. "Energy Production and Consumption". In: *Our World in Data* (2020). https://ourworldindata.org/energy-production-consumption.

[8] Anders SG Andrae and Tomas Edler. "On global electricity usage of communication technology: trends to 2030". In: *Challenges* 6.1 (2015), pp. 117–157.

[9] Nicola Jones et al. "How to stop data centres from gobbling up the world's electricity". In: *Nature* 561.7722 (2018), pp. 163–166.

[10] Hessam Lavi. *Measuring greenhouse gas emissions in data centres: the environmental impact of cloud computing | Insights & Sustainability | Climatiq.* `https://www.climatiq.io/blog/measure-greenhouse-gas-emissions-carbon-data-centres-cloud-computing`. (Accessed on 04/22/2024). Apr. 2022.

[11] Precedence Research. *Data Center Market Size To Hit Around USD 640.77 Bn by 2032.* `https://www.precedenceresearch.com/data-center-market`. (Accessed on 04/24/2024). Dec. 2023.

[12] Eduard Oró and Jaume Salom. "Data centre overview". In: *Advanced Concepts for Renewable Energy Supply of Data Centres.* River Publishers, 2022, pp. 1–26.

[13] Arman Shehabi et al. "United states data center energy usage report". In: (2016).

[14] Miyuru Dayarathna, Yonggang Wen, and Rui Fan. "Data center energy consumption modeling: A survey". In: *IEEE Communications surveys & tutorials* 18.1 (2015), pp. 732–794.

[15] Girish Ghatikar. "Demand response opportunities and enabling technologies for data centers: Findings from field studies". In: (2012).

[16] Facebook. *How we're helping fight climate change.* Aug. 2021. URL: https://about.fb.com/news/2021/06/2020-sustainability-report-how-were-helping-fight-climate-change/.

[17] Lucas Joppa. *Made to measure: Sustainability Commitment Progress and updates.* July 2021. URL: https://blogs.microsoft.com/blog/2021/07/14/made-to-measure-sustainability-commitment-progress-and-updates/.

[18] Ross Koningstein. *We now do more computing where there's cleaner energy.* May 2021. URL: https://blog.google/outreach-initiatives/sustainability/carbon-aware-computing-location/.

[19] Rich Miller. *EBay goes live with its bloom-powered Data Center.* Sept. 2013. URL: https://www.datacenterknowledge.com/archives/2013/09/26/ebay-goes-live-with-its-bloom-powered-data-center#close-modal.

[20] Heather Clancy. *New eBay Data Center runs almost entirely on Bloom Fuel Cells.* Oct. 2013. URL: https://www.forbes.com/sites/heatherclancy/2013/10/02/new-ebay-data-center-runs-almost-entirely-on-bloom-fuel-cells/?sh=5f4029ef60b5.

[21] North C. *Duurzame datacenters: Verklein UW CO2 footprint: Northc.* URL: https://www.northcdatacenters.com/over-northc-datacenters/duurzame-datacenters/.

[22] Noelle Walsh. *Progress on our commitment to sustainable backup power in datacenters by 2030 | Microsoft Azure Blog — azure.microsoft.com.* https://azure.microsoft.com/en-us/blog/progress-on-our-commitment-to-sustainable-backup-power-in-datacenters-by-2030/. [Accessed 12-02-2024].

[23] Peter Judge. *Microsoft builds 3MW hydrogen fuel cell backup power plant.* July 2022. URL: https://www.datacenterdynamics.com/en/news/microsoft-builds-3mw-hydrogen-fuel-cell-backup-power-plant/.

[24] Cristian A Morales Vásquez. "Evaluation of Medium Speed Diesel generator sets and energy storage technologies as alternatives for reducing fuel consumption and exhaust emissions in electric propulsion systems for PSVs". In: *Ship Science and Technology* 9.18 (2016), pp. 49–62.

[25] Y Azoumah, D Yamegueu, and X Py. "Sustainable electricity generation by solar pv/diesel hybrid system without storage for off grids areas". In: *IOP Conference Series: Materials Science and Engineering.* Vol. 29. 1. IOP Publishing. 2012, p. 012012.

[26] D Yamegueu, Y Azoumah, and X Py. "Experimental and economical study of sustainable electricity generation by solar PV/diesel hybrid systems without storage for off-grid areas". In: *Energy and Sustainability* (2011), p. 37.

[27] SK A Shezan. "Optimization and assessment of an off-grid photovoltaic–diesel–battery hybrid sustainable energy system for remote residential applications". In: *Environmental Progress & Sustainable Energy* 38.6 (2019), e13340.

[28] Andy van den Dobbelsteen and Deirdre van Gameren. *Sustainable TU Delft - vision ambition and action plan V5 ...* 2021. URL: https://d2k0ddhflgrk1i.cloudfront.net/TUDelft/Research%20sites/Sustainability/Sustainable%20TU%20Delft%20-%20Vision%20ambition%20and%20action%20plan%20v5.3_220927.pdf.

[29] Deirdre van Gameren and Andy van den Dobbelsteen. "The Sustainable Campus: Working towards a Carbon-Neutral University". In: *Measuring Net Zero: Carbon accounting for buildings and communities.* 2023, pp. 273–285.

[30] Haisheng Chen et al. "Progress in electrical energy storage system: A critical review". In: *Progress in natural science* 19.3 (2009), pp. 291–312.

[31] Marcelo Gustavo Molina. "Dynamic modelling and control design of advanced energy storage for power system applications". In: *Dynamic Modelling* 300 (2010).

[32] Willem Johannes van Egmond. "Concentration Gradient Flow Batteries: salinity gradient energy systems as environmentally benign largescale electricity storage". PhD thesis. Wageningen University and Research, 2018.

[33] Thu-Trang Nguyen et al. "A review on technology maturity of small scale energy storage technologies". In: *Renewable Energy and Environmental Sustainability* 2 (2017), p. 36.

[34] Anders SG Andrae. "Comparison of several simplistic high-level approaches for estimating the global energy and electricity use of ICT networks and data centers". In: *International Journal* 5.51 (2019), pp. 2414–2077.

[35] Charles T Decker et al. "Opportunities for waste heat recovery at contingency bases". In: (2016).

[36] Turnkey Industries. *What Is The Life Expectancy Of Industrial Generators? » Turnkey Industries.* https://turnkey-industries.com/what-is-the-life-expectancy-of-industrial-generators. (Accessed on 04/25/2024). July 2022.

[37] Sean J Ericson and Daniel R Olis. *A comparison of fuel choice for backup generators.* Tech. rep. National Renewable Energy Lab.(NREL), Golden, CO (United States), 2019.

[38] Caterpillar Generators. *Biofuel for CAT generators.* URL: https://www.cat.com/en_US/by-industry/electric-power/electric-power-industries/renewable-liquid-fuels.html.

[39] Martin Pechout et al. "Comparison of hydrogenated vegetable oil and biodiesel effects on combustion, unregulated and regulated gaseous pollutants and DPF regeneration procedure in a Euro6 car". In: *Science of the Total Environment* 696 (2019), p. 133748.

[40] Oliver Schmidt et al. "Future cost and performance of water electrolysis: An expert elicitation study". In: *International journal of hydrogen energy* 42.52 (2017), pp. 30470–30492.

[41] Raluca-Andreea Felseghi et al. "Hydrogen fuel cell technology for the sustainable future of stationary applications". In: *Energies* 12.23 (2019), p. 4593.

[42] clean energy institute. *Lithium-ion battery - clean energy institute.* Jan. 2024. URL: https://www.cei.washington.edu/research/energy-storage/lithium-ion-battery/#:~:text=In%20part%20because%20of%20lithium%E2%80%99s,unit%20mass%20and%20unit%20volume..

[43] Dragonfly Energy. *A guide to the 6 main types of lithium batteries.* Dec. 2022. URL: https://dragonflyenergy.com/types-of-lithium-batteries-guide/#:~:text=Lithium%20Cobalt%20Oxide,power%20over%20a%20long%20period..

[44] Green economy media. *Nas batteries: Superior Technology for long-duration stationary energy storage.* Aug. 2020. URL: https://greeneconomy.media/nas-batteries-superior-technology-for-long-duration-stationary-energy-storage/.

[45] Research and Blackridge Consulting. *Here's what you need to know about sodium sulfur (NAS) batteries.* URL: https://www.blackridgeresearch.com/blog/heres-what-you-need-to-know-about-advantages-disadvantages-applications-of-sodium-sulfur-sulphur-nas-batteries.

[46] Paul Denholm and Gerald L Kulcinski. "Life cycle energy requirements and greenhouse gas emissions from large scale energy storage systems". In: *Energy Conversion and Management* 45.13-14 (2004), pp. 2153–2172.

[47] Olusola Fajinmi et al. "Compressed Air Energy Storage as a Battery Energy Storage System for Various Application Domains: A Review". In: *Energies* 16.18 (2023), p. 6653.

[48] Jihong Wang et al. "Compressed Air Energy Storage". In: *Energy Storage.* World Scientific, 2017, pp. 81–115.

[49] Kai Xu et al. "A Review of Flywheel Energy Storage System Technologies". In: *Energies* 16.18 (2023), p. 6462.

[50] Rafael Peña-Alzola et al. "Review of flywheel based energy storage systems". In: *2011 International Conference on Power Engineering, Energy and Electrical Drives.* IEEE. 2011, pp. 1–6.

[51] Erik A Alsema. *Environmental life cycle assessment of solar home systems.* Department of Science Technology and Society, Utrecht University Utrecht ..., 2000.

[52] Mark Goedkoop, Renilde Spriensma, et al. *The eco-indicator 95.* PRé Consultants Amersfoort, the Netherlands, 1995.

[53] Volker Quaschning und Bernhard Siegel. *Specific carbon dioxide emissions of various fuels.* `https://www.volker-quaschning.de/datserv/CO2-spez/index_e.php`. (Accessed on 04/28/2024). Nov. 2022.

[54] Kelly Benton, Xufei Yang, and Zhichao Wang. "Life cycle energy assessment of a standby diesel generator set". In: *Journal of Cleaner Production* 149 (2017), pp. 265–274.

[55] Miguel Brandão, Reinout Heijungs, and Annette R Cowie. "On quantifying sources of uncertainty in the carbon footprint of biofuels: crop/feedstock, LCA modelling approach, land-use change, and GHG metrics". In: *Biofuel Research Journal* 9.2 (2022), pp. 1608–1616.

[56] Miguel Brandão et al. "The modelling approach determines the carbon footprint of biofuels: the role of LCA in informing decision makers in government and industry". In: *Cleaner Environmental Systems* 2 (2021), p. 100027.

[57] L Oliveira et al. "Environmental performance of electricity storage systems for grid applications, a life cycle approach". In: *Energy conversion and management* 101 (2015), pp. 326–335.

[58] Peter Van den Bossche et al. "SUBAT: An assessment of sustainable battery technology". In: *Journal of power sources* 162.2 (2006), pp. 913–919.

[59] JL Sullivan and Leigh Gaines. "A review of battery life-cycle analysis: state of knowledge and critical needs." In: (2010).

[60] Md Mustafizur Rahman et al. "Energy and environmental footprints of flywheels for utility-scale energy storage applications". In: *e-Prime-Advances in Electrical Engineering, Electronics and Energy* 1 (2021), p. 100020.

[61] Md Mustafizur Rahman. "Economic and Environmental Assessment of Large-scale Electro-chemical and Flywheel Energy Storage Systems for Stationary Applications". In: (2022).

[62] Ember. *Electricity price Netherlands 2019 - 2024.* Mar. 2024. URL: `https://www.statista.com/statistics/1314549/netherlands-monthly-wholesale-electricity-price/`.

[63] Tango. *Historiek Adviesprijzen.* URL: `https://www.tango.nl/historiek-adviesprijzen`.

[64] Riccardo Amirante et al. "Overview on recent developments in energy storage: Mechanical, electrochemical and hydrogen technologies". In: *Energy Conversion and Management* 132 (2017), pp. 372–387.

[65] S Ould Amrouche et al. "Overview of energy storage in renewable energy systems". In: *International journal of hydrogen energy* 41.45 (2016), pp. 20914–20927.

[66] Dale AC Brownson, Dimitrios K Kampouris, and Craig E Banks. "An overview of graphene in energy production and storage applications". In: *Journal of Power Sources* 196.11 (2011), pp. 4873–4885.

[67] Verena Jülch. "Comparison of electricity storage options using levelized cost of storage (LCOS) method". In: *Applied energy* 183 (2016), pp. 1594–1606.

[68] Xing Luo et al. "Overview of current development in electrical energy storage technologies and the application potential in power system operation". In: *Applied energy* 137 (2015), pp. 511–536.

[69] Behnam Zakeri and Sanna Syri. "Electrical energy storage systems: A comparative life cycle cost analysis". In: *Renewable and sustainable energy reviews* 42 (2015), pp. 569–596.

[70] Ahmed F Zobaa et al. *Energy storage at different voltage levels: technology, integration, and market aspects.* Vol. 111. Energy Engineering, 2018.

Journal of Physics: Conference Series **2929** (2024) 012009 doi:10.1088/1742-6596/2929/1/012009

Digital Dynamic Agent (DDA) Driven Energy Contract Negotiation

Milad Rezamand[1*], Rupp Carriveau[1], Jacqueline Stagner[1] and Matt Davison[2]

[1]Environmental Energy Institute, University of Windsor, Windsor, Canada
[2] Department of Statistical and Actuarial Sciences, Western University, London, Canada

[*]E-mail: rezaman@uwindsor.ca

Abstract. Digital Dynamic Agent (DDA) introduces the new paradigm of digital dynamic agent-driven energy contract negotiation, which is dynamic and real-time in adjustment. Long-term static contracts normally produce one-sided traditional market outcomes. On the contrary, DDA driving will be continuous in nature, at a level of rapid negotiations over the underlying transactions. This paper proposes a new generic DDA framework that negotiates energy contracts between grid agents and renewable generation sources, especially wind farms equipped with battery storage. The agents dynamically update their strategy concerning real-time market conditions, demand, and storage capacity. By simulation over one week with real data from the wind farms, it is evident that the DDA framework is better adapted in using battery storage for significantly lesser wastage of energy. Interestingly, the system secured energy at competitive prices in both live energy and capacity markets; it showed an improvement in cost savings compared to conventional static contracts. It also allows for effective energy procurement within diverse market states, which enables greater flexibility and responsiveness in energy management through the optimization of the negotiation process. The current paper highlights the capability of the framework to enhance negotiation outcomes while allowing the broader transition toward renewable energy markets.

1. Introduction

The transition to renewable energy sources forms significant challenges in aligning supply with demand. Impotent traditional energy markets in handling the dynamic nature of renewables makes innovative approaches necessary in contract negotiations. A Digital Dynamic Agent (DDA) system is presented in this paper to enable autonomous negotiations between grid agents and wind energy providers by modelling battery storage and negotiator to simulate realistic negotiation scenarios.

1.1 Literature Review

The digital agent concept in negotiation approaches is not entirely new; however, its application within the energy sector has gained significant attention only in recent years. Early studies by Jennings et al. [1] investigated the potential of intelligent agents in electronic marketplaces. The study helped lay the groundwork for their use in various negotiation scenarios. Research by Faratin et al. [2] explored the algorithms and mechanisms that facilitate these agents to perform complex negotiations autonomously. They highlighted their enhanced decision-making capabilities. Bazzan and Klügl [3] examined agent-based modeling potentials for simulating market behaviors and optimizing energy distribution. Afterward, Vytelingum et al. [4] introduced autonomous trading agents capable of participating in dynamic pricing markets. They emphasized agents' benefits in reducing operational costs and improving market efficiency.

Energy markets present unique challenges including the dynamic nature of energy supply and demand, unstable prices, and regulatory constraints. To address this, Ramchurn et al. [5] showed how DDAs could optimize energy negotiations by expecting market trends and consumer behavior. Further progress by Shafie-khah et al. [6] and Aitzhan and Svetinovic [7] focused on blockchain technology development. It improved the DDA security and transparency.

More recent contributions by Zhang et al. [8] and Aklilu and Ding [9] have shown the role of machine learning and big data analytics in the DDA predictive capabilities. They have delivered better forecasts of market fluctuations, helping negotiate more promising terms in the interests of their human counterparts. Besides, Zhang et al. [10] developed a multi-agent system to emphasize the collaborative potential of DDAs. The proposed method contains multiple agents that negotiate to achieve mutually beneficial outcomes.

Despite these improvements, the practical implementation of DDAs in energy contract negotiations remains in its early stages. Challenges such as regulatory compliance, integration with existing infrastructure, and securing stakeholder acceptance require further investigation. However, the potential benefits of DDAs, including increased efficiency, cost savings, and enhanced market responsiveness warrant future research and development.

1.2 Research Objectives

This paper brings several important contributions to the field of energy market dynamics and contract negotiations:

- A new DDA framework is proposed for energy contract negotiations to bridge a gap in the existing literature. It allows contracts to adapt dynamically to real-time data and market conditions by simulating scenarios with wind farms and battery storage systems.
- The DDA framework helps make negotiations more efficient and effective, which contributes to overall market stability.
- Our DDA framework lays the foundation for future studies into more complex negotiation scenarios by integrating various renewable energy sources.

2. A Preliminary theory of the proposed DDA method

The proposed Digital Dynamic Agent (DDA) framework consists of Grid Agents and wind energy providers. As illustrated in Figure 1, these agents negotiate energy contracts based on current demand, generation capacity, storage availability, and utility function evaluation.

In each negotiation round, agents begin by verifying if their negotiation deadlines have not been reached and if the available energy and storage capacity satisfy the demand. If these criteria are satisfied, agents proceed to adjust their target utilities based on their concession strategies. They then apply utility functions to determine the payoff of an offer. Depending on the outcome of this calculation, agents will either accept the offer or reject it and present a counteroffer.

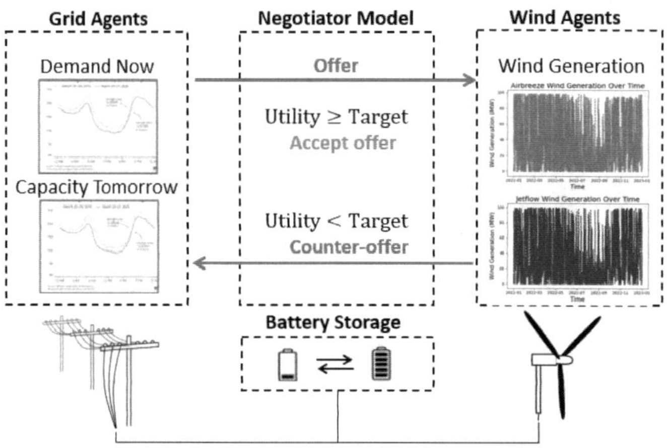

Figure 1. The Proposed Autonomous Negotiators

2.1 Battery storage model

Battery storage is managed through charging and discharging functions based on current wind generation and demand as shown in Algorithm 1, ensuring that energy is efficiently stored and utilized.

Algorithm 1. Battery Storage Model

```
input current wind generation data (WG_t^j)
avlEn_t^j = WG_t^j + BS_t^j
if avlEn_t^j ≥ D_t^{j'} then
    excEn_t^j = WG_t^j − D_t^{j'}
    if excEn_t^j ≥ 0 then
        BS_t^j = +excEn_t^j
    Else
        BS_t^j = −excEn_t^j
    end if
end if
```

[a] Where t is the negotiation time. $avlEn_t^j$, WG_t^j, and BS_t^j denote the available energy, wind generation, and battery storage of agent j. $D_t^{j'}$ denotes the demand of agent j' and $excEn_t^j$ represents the excess energy.

In each round of negotiations (t), the available energy of agent j is calculated by adding its wind generation of the agent j and battery storage. If the available energy meets the demand of agent j', an offer is proposed, and the negotiation model presented in Algorithm *2* is executed. The excess energy is then utilised to charge battery. This process ensures that any surplus energy is efficiently stored for future use, enhancing the overall reliability and stability of the energy supply. By dynamically adjusting to real-time data, the system can optimize energy utilization and minimize waste. Additionally, this method promotes a more sustainable energy market by effectively integrating renewable energy sources into the grid. Overall, the approach described allows for more flexible and responsive energy management, which is crucial in addressing the fluctuating nature of renewable energy generation.

The DDA framework uses a battery storage model, which operates the flow of energy by either charging or discharging. However, efficiency factors and limit checks should be considered in them for its closeness to reality. Fundamentally, in practical scenarios, there are usually certain efficiency losses in battery storage systems, generally ranging between 10% and 20%, caused by conversions of electricity in the time of charging and discharging. An efficiency factor is included for both the charging and discharging processes in the model in order to simulate it more realistically.

Furthermore, limit checks are placed to avoid overcharging and deep discharging conditions of the batteries, which have detrimental effects on its performance. The capacity of the storage system is limited, and it accepts energy only up to that limit. Similarly, it stops discharging when the battery reaches its minimum allowable charge level. With this addition, the DDA model not only maximizes the energy utilization but also agrees within the operational constraints of battery systems.

2.2 DDA model

The negotiation process involves proposing prices within predefined ranges and accepting offers based utility evaluation, provided in Algorithm 2. By dynamically proposing prices, agents can adapt to changing market demands and competitive landscapes. Utility evaluation plays a crucial role in this model, as it assesses the value of received offers by considering factors such as energy availability, price, delivery time, and target utility. This model facilitates efficient and effective negotiations in digital environments, making it applicable across diverse industries such as e-commerce, automated trading, and supply chain management.

In each negotiation round (t), an offer is proposed to agent j by agent j'. If the negotiation time is within the prescribed negotiation deadline, the agent j adjusts its target utility and

Algorithm 2. DDA Model

initial offer $x_t^{j'\to j}$ based on energy availability
If $t \le t_{Max}$ then
  concede target utility using the following equation.

$$T_t^j = T_{Max}^j - (T_{Max}^j - T_{Min}^j)(\frac{t}{t_{Max}})^\alpha \quad (1)$$

  calculate utility function using the following equation.

$$U^j(x_t^{j'\to j}) = \sum_k w_k^j E_k^j(x_t^{j'\to j}[k]) \quad (2)$$

  if $U^j(x_t^{j'\to j}) \ge T_t^j$ then
    accept offer $x_{i'\to i}^t$
  Else
    learn and update the estimation on opponent j' preference
    propose counteroffer $x_t^{j\to j'}$
  end if
Else
  quit negotiation
end if

[a] Where $x_t^{j'\to j}$ is the offer proposed by agent j' to agent j, t_{Max} denotes the negotiation deadline, and T_t^j represents the target utility of agent j. T_{Min}^j and T_{Max}^j denote the minimum and maximum target utility of agent j. The value of α is used to change the concession strategy. E^j and U^j are the evaluation and utility functions used by agent j to evaluate offer $x_t^{j'\to j}$ proposed by agent j'. k denote one of the issues that are negotiated between the two agents and w_k^j represents the issue k weighting of agent j.

evaluates the offer based on its utility assessment and the conceded target. Should the agent j find the offer unacceptable, it formulates a counteroffer directed to the agent j', considering the agent j' preferences.

Utility functions and concession strategies form the basis in the DDA negotiation framework. Utility functions model the agent's preferences about different offers, returning a numeric desirability value for an offer, depending on multiple factors—energy availability, price, delivery time, and storage capacity.

The utility function (U_j) for an agent j is defined as:

$$U_j = w_1 \times Price + w_2 \times DeliveryTime + w_3 \times EnergyAvailability + w_4 \times BatteryStorage$$

Where w_1, w_2, w_3, and w_4 are the weights assigned to each factor, which can vary based on the agent's priorities. These weights allow the placing of greater importance on one aspect of the negotiation, such as low price or immediate delivery, over another due to the situation.

The concession strategy allows regulating how an agent will update its target utility throughout the course of the negotiation process. This model uses a time-based strategy in its concession strategy, whereby around the time it is getting near the deadline for the negotiation, the agent becomes willing to give way more and more. This is reflected in the concession parameter α, which determines the speed at which an agent will concede utility during later iterations of the negotiation: the higher the α value, the more aggressive the agent in making concessions; inversely, the lower α corresponds to a more reserved negotiation stance. For example, in this work, an α value of 2 means that the agent will make small concessions at the beginning but will have more considerable concessions toward the end to make sure that the excess energy is sold by the end of the negotiation window.

Notably, The agents in the DDA framework are characterized by a set of parameters that govern their behavior in negotiations. These parameters include:

- Target Utility: This is the level of utility that an agent would like to achieve during the process of negotiation.
- Concession Rate (α): Controls the rate at which the agent is ready to concede its target utility over time.
- Oppositional learning: refers to the process in which agents estimate and adapt to the opponent's strategy based on some historical negotiation data and predictive modeling.

The agents learn the estimation of opponents and update that by analyzing the patterns in structure and opponent's response over time. This is achieved through a sort of Bayesian learning and neural networks where agents use the feedback from previous rounds to predict their opponent's concession behavior. This is evident when, for example, an agent finds out that the opponent always concedes in just a few initial rounds of negotiations; the former will therefore automatically review its concession strategy with the aim of achieving the best possible outcome. Agents, over successive rounds of negotiation, refine their estimations using historical data on offers and readjust concession strategies. This adaptiveness now enhances the negotiation process by using the capability of the agents to make more accurate counter-offers while learning the opponents' preference and tendencies.

Journal of Physics: Conference Series **2929** (2024) 012009 doi:10.1088/1742-6596/2929/1/012009

3. Design studies and experimental results

This section introduces the configuration of the proposed autonomous negotiation agent and analyses experimental results to assess the performance of the proposed method.

3.1 Experimental case studies

Figure 2 depicts the Grid Agent's interest in securing energy for a Live Energy Market and future energy for a Capacity Market. The Grid Agent includes both a Central Grid with a 10 MWh live energy demand every 10 minutes, paying \$70-\$110 per MWh, and a Local Grid with a 10 MWh capacity demand, also paying \$70-\$110 per MWh.

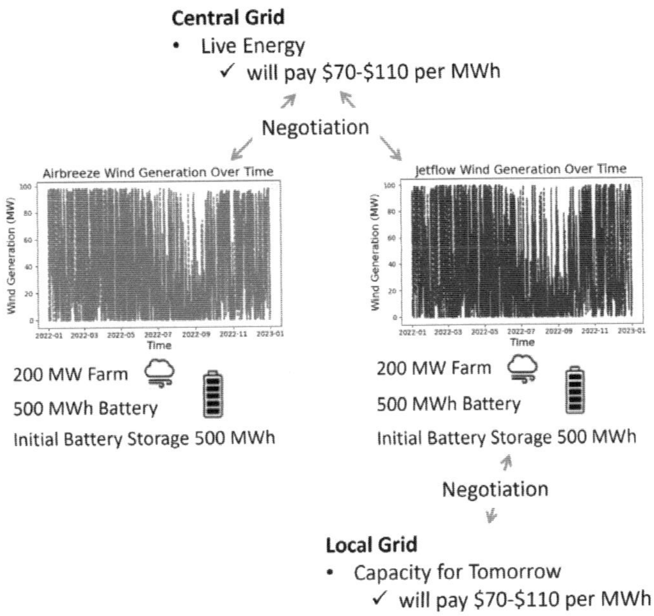

Figure 2. Experimental case study

The case study is based on data from 89 turbines from two different commercial 200 MW wind farm sites with 500 MWh batteries, Airbreeze and Jetflow, located in Southern Ontario, Canada. For this analysis, we've used 10-minute SCADA data recorded from 2022 to 2023. At the beginning of the negotiations, Airbreeze's and Jetflow's batteries are full.

3.2 Design implementation

In this study, 10-minute experimental wind generation data are used to model a Digital Dynamic Agent (DDA). The autonomous negotiation between grid and wind agents occurs every 10 minutes over a period of 7 days, incorporating wind battery storage and a negotiation model.

If the available energy from wind generation and battery storage meets the demand, the grid agent proposes an offer to the wind agent. If the negotiation time is within the deadline, the wind agent concedes its target utility using Equation 1. An α value of 2 is used, indicating that the wind agent will make small concessions at the beginning of the negotiation, but increase the concession degree in later rounds. This strategy ensures that the excess energy can be sold by the end of the negotiation period. The wind agent evaluates the offer based on its utility assessment and accepts the offer if it meets the conceded target. If it does not, the excess energy is stored in the battery, as illustrated in Algorithm 1, and the wind farm presents a counteroffer to the grid agent. This iterative process continues until an agreement is reached or the negotiation time exceeds the deadline. During each round, both agents make strategic adjustments to their offers, reflecting changes in market conditions and internal constraints. The negotiation dynamics are further

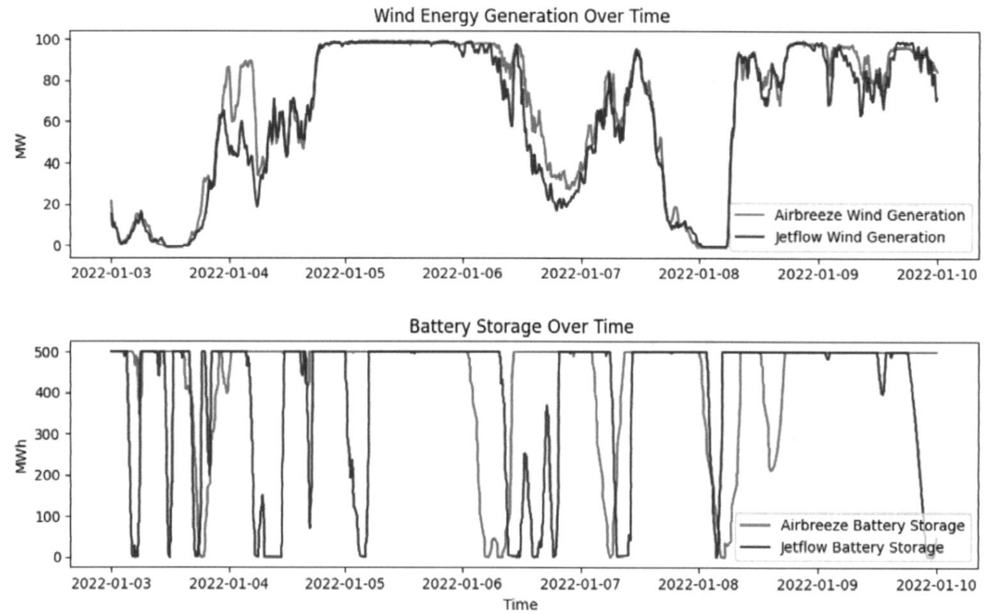

Figure 3. Wind generation and battery storage over negotiation time

influenced by the agents' historical interactions and predictive modelling of opponent behavior, enhancing the sophistication of the counteroffers.

Additionally, each agent's decision-making process incorporates a blend of concession strategies and aspiration adjustments, striving to maximize their respective utilities while maintaining a feasible compromise. The agents utilize advanced algorithms to analyze potential outcomes and iteratively refine their negotiation tactics.

3.3 Experimental results

Figure *3* presents the wind generation data for the Airbreeze and Jetflow wind farms, along with the charge and discharge activities of their battery storage systems during a seven-day negotiation period.

The simulation outputs the accepted offers for live energy and capacity markets, including the prices accepted over the negotiation period. Figure 4 depicts the negotiation offers and counteroffers between wind and grid agents. The circles in blue and red represent the offers made

Journal of Physics: Conference Series **2929** (2024) 012009 doi:10.1088/1742-6596/2929/1/012009

to Jetflow and Airbreeze by grid agents, respectively. If an offer is rejected, indicated by black

Negotiations on Jan 4th at 00:00	Utility	Target	Decision
Offer to Jetflow	0.08	0.89	Rejected by Jetflow
Counteroffer by Jetflow	0.994	0.94	Rejected by Grid
Offer to Airbreeze	0.717	0.89	Rejected by Airbreeze
Counteroffer by Airbreeze	0.93	0.94	Accepted by Grid

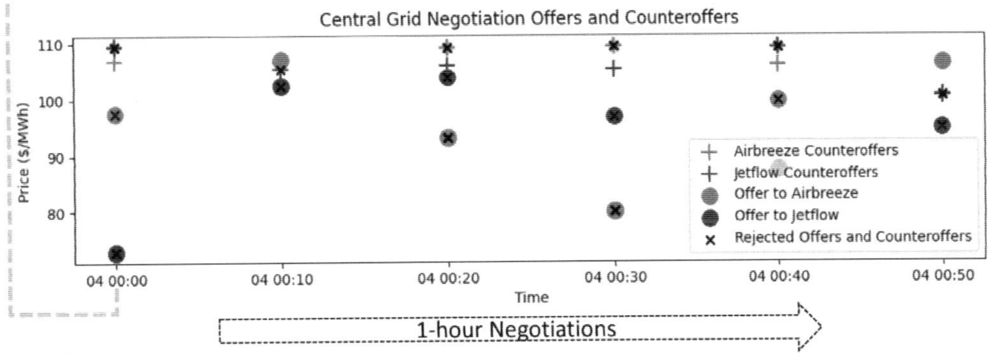

Figure 4. Offers and counteroffers between Central Grid and Wind Farms

crossed, a counteroffer is proposed by wind agents to grid agents, illustrated by plus signs.

Figure 4 illustrates a 1-hour negotiation that begins with an offer from Central Grid to Jetflow on January 4th at 00:00. Jetflow rejected the offer based on its utility evaluation and proposed a counteroffer to Central Grid. The grid agent then rejected this counteroffer and made an offer to Airbreeze. Airbreeze also rejected the offer based on its utility evaluation and proposed a counteroffer to Central Grid, which the grid accepted. This negotiation process continues, with agents gradually conceding their targets to reach a mutual agreement, resulting in Figure 5 and Figure 6. Additionally, Figure 4 highlights the calculated utilities of offers and counteroffers compared to the target utilities. Notably, the criteria for a wind agent to accept an offer is that the utility exceeds their target, while the grid agent accepts a counteroffer when the utility is below its target.

Figure 5 and Figure 6 illustrate the accepted offers for live energy between the Central Grid and wind farms, and for capacity demands between the Local Grid and wind farms, respectively.

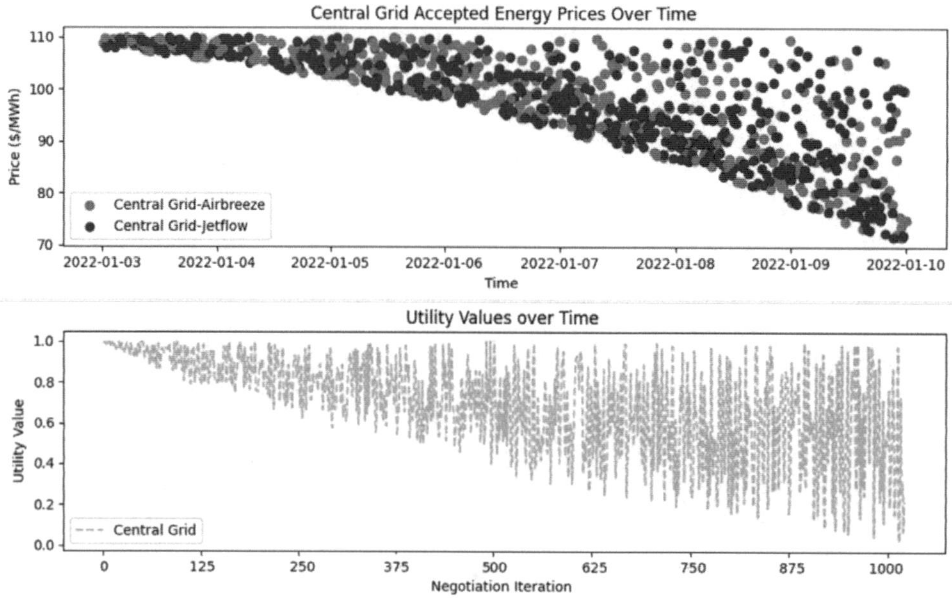

Figure 5. Central grid accepted negotiations

At the outset of negotiations, agents typically enter into modest agreements. However, as the deadline approaches, the intensity and frequency of negotiations generally escalate, with both parties striving to reach a mutually advantageous settlement.

This dynamic is evident from the utility curve over time, which normalizes the negotiated price within the acceptable price range, providing a consistent metric for evaluating the quality of the deals. Initially, the negotiated price must equal the maximum target price to be accepted, ensuring that the utility equals one. Nevertheless, as the negotiation period progresses, wind companies exhibit a willingness to concede their target utility to ensure the sale of excess energy by the conclusion of the seven-day negotiation window.

This strategic concession allows negotiator agents from both the central and local grids to secure agreements that are closer to their preferred price points as the negotiation deadline looms. Consequently, lower utility values generally signify more successful negotiations, as they reflect the agents' ability to close deals that are more favorable to their initial pricing objectives. This means throughout the process of negotiation, agents may have to concede at one point in order to settle the agreement. Thus, a lower utility value means the agent accepts an offer farther from its target; therefore it has given way a lot to seal the deal.

From the market perspective, a successful negotiation is one in which both parties agreed on a price and terms within a limited timeframe, thus avoiding the inefficiencies of failed negotiations. While the lower utility values indicate that an agent has settled for more than it has bargained for, the completion of the transaction is considered healthy, especially in situations where unsold energy might be wasted.

Success," however, is a relative term: in energy markets, surplus energy storage is limited, and hence successfully closing a deal-even at lower utility value-prevents the loss of potential revenue. The lower utility values can thus be seen as the willingness to close deals near negotiation deadline and hence as marker of successful negotiation closure.

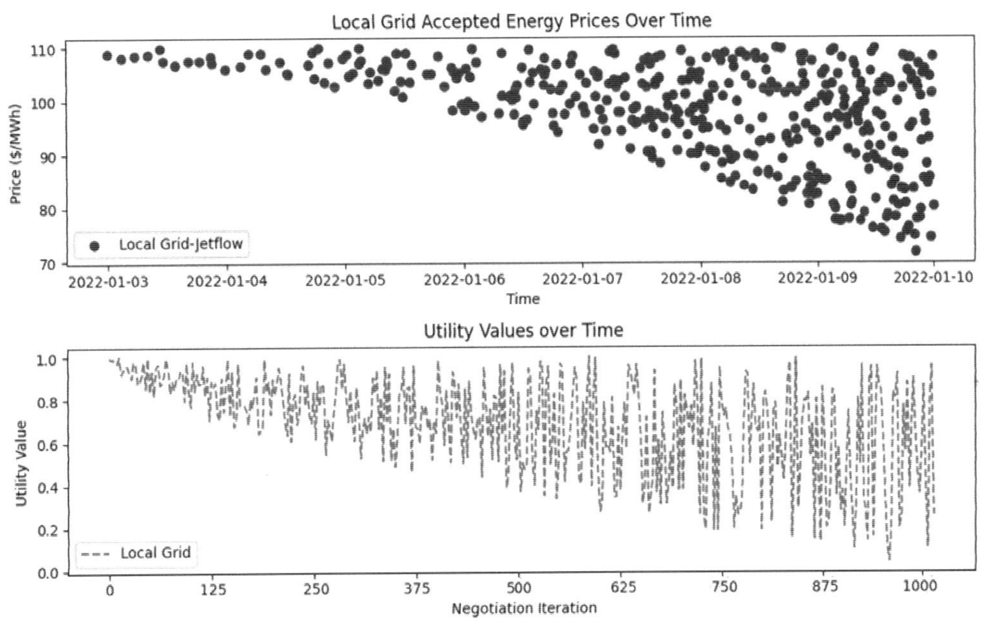

Figure 6. Local grid accepted negotiations

As shown in Figure 7, if the DDA approach is to provide competitive pricing, it must be compared with a conventional static (Human) contract scenario. Conventional energy contracts are usually based on a preestimation price over a long period and may not represent the market price at that particular time. In such situations, the market could overvalue or undervalue energy. For example:

- In periods of overproduction, static contracts might commit producers to low rates of sale, while the DDA system could permit dynamic price increases as determined by real-time demand.
- On the other hand, when energy undersupply occurs, static contracts would make grid operators overpay for it, whereas the DDA system should be able to bargain more advantageous terms or adjust procurement strategies dynamically.

The competitiveness of DDA-driven negotiations can be gauged by comparing revenue generated in DDA scenarios with conventional fixed-rate contracts under similar market conditions. Key metrics would include:

- Responsiveness of price fluctuation: The time it takes for the system to respond to changes in supply and demand.
- Cost saving: A comparison of the long-term costs incurred by grid operators and energy producers in DDA-driven contracts versus static ones.

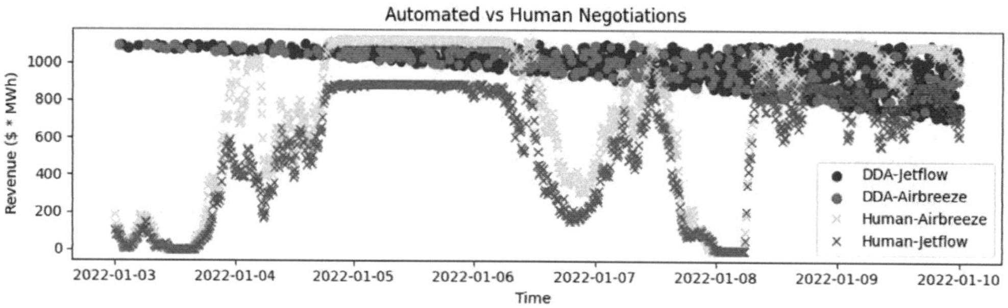

Figure 7. Revenue Competitiveness: Comparison to Conventional Scenarios

- Efficiency gains: Measuring how much more energy is efficiently distributed (e.g., through battery storage) versus wasted due to static contract limitations.

This sets up some enlightening comparisons-those of averages in energy prices or volatility, for example-which demonstrate how the system of DDA ensures revenue competitiveness with flexibility in balance with market realities.

4. Conclusion

This paper presents a DDA-driven framework for dynamic energy contract negotiations, incorporating wind energy generation and battery storage models. The simulation results validate the framework's ability to optimize energy procurement and manage costs effectively. Future work will focus on enhancing the negotiation algorithms and integrating additional renewable energy sources.

The simulation results demonstrate the efficacy of the DDA framework in managing dynamic energy contract negotiations. The central grid successfully secures energy at competitive prices, while the local grid benefits from consistent capacity procurement. The inclusion of wind energy and battery storage models provides a realistic representation of renewable energy dynamics.

Furthermore, the temporal aspect of negotiation dynamics underscores the importance of strategic patience and tactical flexibility. Wind companies, anticipating the approaching deadline, adjust their pricing strategies to align with market realities and grid requirements. This adaptive approach not only facilitates the sale of surplus energy but also enhances the overall efficiency and stability of the energy market. The evolving negotiation landscape, marked by fluctuating utility values and intensified bargaining efforts, highlights the intricate balance between immediate gains and long-term strategic positioning in the energy trading ecosystem.

4.1 Limitations and Future Challenges

Although the proposed DDA-driven framework presents a huge potential for modernizing the way energy trading is negotiated, there are several limitations and challenges that should be considered in future research and real-world implementation.

- Regulatory and Policy Barriers: The introduction of the mechanisms of DDA in energy markets should be rigorously performed in a regulating and policy-guided environment. They have to operate within existing legal frameworks that control the purchase and grid management of energy. Those may also be viewed as deployment challenges, taking into consideration how far these regulations can vary from region to region, especially across border energy trading.
- Infrastructure Compatibility: Just an integration bottleneck exists currently with the DDA system for energy trading infrastructures. Large portions of the energy market systems, with big parts designed for static, long-term contracts, may accordingly need substantial reconfigurations to support continuous, real-time negotiations. Enabling seamless communications between grid agents and renewable energy providers may require a major technological upgrade for high-speed, secure data exchange.
- Scalability: Although the paper illustrates the DDA framework within a controlled, small-scale simulation with wind farms and battery storage participating, how this would scale remains very much an open question. Such large-scale deployment across a. set of energy producers, consumers, and storage providers may add more complexities to the system, such as congestion in negotiation traffic, delays in decision-making, and computational burden due to multiple parallel negotiations.
- Security and privacy are another critical limitations that are to be defined by the DDA mechanism, especially while digitization is performed in energy trading. Blockchain and cryptography are proposed, but scaling this up brings with it both technical and cost-related challenges. It is a challenge to guarantee that all transactions are secure without impeding on the speed and efficiency of the negotiation process.
- Energy Market Volatility: The renewable energy markets are usually very volatile, considering the ever-changing supply from sources like wind and solar. While these dynamics are indeed possible under the DDA system through models which incorporate battery storage and allow flexible negotiation processes, the performance in extreme market conditions-for instance, during sustained periods of low renewable energy generation or sudden spikes in demand-is not fully explored to date. Future research will focus on the robustness of the DDA framework under these conditions.

References

[1] Jennings N R, Faratin P., Lomuscio A R, Parsons S, Sierra C and Wooldridge M 2001 Automated negotiation: prospects, methods and challenges International Journal of Group Decision and Negotiation **10** 199-215

[2] Faratin P, Sierra C and Jennings N R 2002 Using similarity criteria to make issue trade-offs in automated negotiations artificial Intelligence **142** 205-237

[3] Bazzan A L C and Klügl F 2005 Introduction to intelligent systems in traffic and transportation *Synthesis Lectures on Artificial Intelligence and Machine Learning* **3** 1-137

[4] Vytelingum P, Ramchurn S D, Voice T D, Rogers A and Jennings N R 2010 Trading agents for the smart electricity grid *9th International Conference on Autonomous Agents and Multiagent Systems*

[5] Ramchurn S D, Vytelingum P, Rogers A and Jennings N R 2012 Putting the 'smarts' into the smart grid: A grand challenge for artificial intelligence *Communications of the ACM* **55** 86-97

[6] Shafie-khah M, Mahmoudi M and Catalão J P S 2016 Optimal trading of plug-in electric vehicle aggregation agents in a market environment for sustainability *Applied Energy* **162** 601-612

[7] Aitzhan N Z and Svetinovic D 2018 Security and privacy in decentralized energy trading through multi-signatures, blockchain and anonymous messaging streams *IEEE Transactions on Dependable and Secure Computing* **15** 840-852

[8] Zhang C, Wu J, Zhou Y, Cheng M and Long C 2020 Peer-to-peer energy trading in a microgrid *Applied Energy* **220** 1-12

[9] Aklilu Y T and Ding J 2021 Survey on Blockchain for Smart Grid Management, Control, and Operation *Energies* **15** 193

[10] Zhang J, Ren F and Zhang M 2016 Prediction of the opponent's preference in bilateral multi-issue negotiation through Bayesian learning *Recent Advances in Agent-based Complex Automated Negotiation* 3-20

Journal of Physics: Conference Series **2929** (2024) 012010 doi:10.1088/1742-6596/2929/1/012010

Numerical Analysis of Pullout Capacity of Composite Suction Caisson Foundation Used For Large Scale Underwater Compressed Gas Energy Storage

***Hu Wang*[1]*, Wei Xiong*[1]*, Zecheng Zhao*[1]*, Tonio Sant*[2]*, Rupp Carriveau*[3]*, David S-K Ting*[3]* and Zhiwen Wang*[1*]**

[1] *Department of Mechanical Engineering, Dalian Maritime University, 116026, Dalian, China*

[2] *Department of Mechanical Engineering, University of Malta, Msida MSD2080, Malta*

[3] *Turbulence and Energy Laboratory, University of Windsor, ON N9B 3P4, Windsor, Canada*

*E-mail: wzw@dlmu.edu.cn

Abstract. The suction caisson foundation, characterized by its unique installation methodology and excellent bearing capabilities, has emerged as a crucial supporting structure for marine constructions. However, the distinctive alternating cyclic high load patterns exhibited by underwater compressed gas energy storage systems significantly challenge the vertical pullout bearing capacity of regular suction caisson foundations. In this study, a novel composite suction caisson foundation is proposed to enhance the vertical pullout capacity. The pullout capacity under perfectly drained conditions is investigated by finite element analysis with ABAQUS/Standard 2020. A comparative analysis of the pullout capacity between regular suction caisson (RSC) and composite suction caisson (CSC) is conducted, focusing on aspect ratios of 0.5, 1, and 1.5. The results show that the pullout load of the proposed CSC, with the same aspect ratio, increased by a range of 12.8% to 100% compared to that of the RSC. As the aspect ratio increases, the trend of increasing pullout load decreases. The variation of displacement for the caisson at its ultimate pullout capability is relatively small, between 0.01D and 0.027D.

1 Introduction

Suction caissons are prevalent foundation structures in marine engineering and deep-water facilities, primarily designed to withstand horizontal and vertical tensile loads. Typically, suction caissons have a hollow cylindrical form with an open bottom and closed top, enabling them to attach to the seabed via a combination of their gravity and internally generated negative pressure. The installation process of suction caissons leverages water pressure differences, specifically by creating negative pressure within the structure, to smoothly penetrate the soil and achieve the desired depth.

In recent years, the utilization of suction caisson foundations has significantly expanded due to their versatility and adaptability to diverse offshore environments. Suction caisson foundations are ubiquitous in supporting a range of offshore facilities, notably offshore wind turbines, offshore platforms, masts, and mooring systems [1,2]. The diverse applicable scenarios inherently dictate the unique types of loads that the suction caisson foundations must endure. For offshore wind turbines, the primary challenge lies in resisting horizontal cyclic loads (control loads) and overturning moments imparted by wind, waves, and currents [3,4]. These loads are highly dynamic and cyclic, posing significant demands on the foundation's lateral stiffness and resistance to fatigue. The suction caisson foundation, with its deep penetration into the seabed and substantial lateral bearing capacity, effectively distributes and dissipates these horizontal and moment loads, ensuring the stability and longevity of the turbine structure. In contrast, the suction caisson foundation's role in submerged floating tunnels (SFTs) primarily involves withstanding low-frequency oblique cyclic loads [5,6]. These loads arise from the complex interplay of water currents and tides, which impart oblique forces on the tunnel structure. The suction caisson foundation, by its ability to transfer these loads efficiently to the surrounding soil, ensures the tunnel's stability and continued operational integrity even under adverse conditions. Notably, the loading patterns in SFTs are more gradual and less frequent compared to those in offshore wind turbines, necessitating a different focus on long-term durability and resistance to gradual deformation.

In the context of underwater compressed gas energy storage systems [7], the suction caisson foundation exhibits distinct characteristics. Here, the foundation primarily bears the significant load of alternating buoyancy caused by the cyclic inflation and deflation of the storage tanks in the vertical direction [8]. This buoyancy load acts as a control load, with its vertical upward force being particularly pronounced. Unlike the horizontal loads encountered in offshore wind turbines or the oblique loads in SFTs, the buoyancy load requires the suction caisson foundation to possess exceptional pullout resistance and vertical stability. Therefore, the pullout capacity of the suction caisson is significant.

Currently, it is recognized that there are three failure modes for vertically pulled suction caissons as shown in Figure 1 [9-11]. The failure mode is closely related to the extraction speed and drainage conditions. The first is frictional failure mode (perfectly drained condition) [12]. Under perfectly drained conditions, the pullout resistance of the suction caisson primarily arises from two sources: the self-weight of the caisson itself and the skin friction acting along the inner and outer skirt perimeters. As the caisson is lifted, a gap gradually forms between the caisson lid and the soil surface due to the dissipation of pore water pressures. The frictional failure mode is characterized by a gradual mobilization of soil resistance along the caisson perimeter, with a relatively low peak resistance compared to other modes. The second is the reverse end bearing failure mode (perfectly undrained condition) [13]. In contrast, under perfectly undrained conditions, suction pressures within the caisson prevent the formation of a gap between the caisson lid and the soil surface. The suction pressure under the lid is one of the main factors contributing to the pullout resistance and is influenced by the pullout rate, loading rate, drainage length and hydraulic conductivity. Suction pressure is defined as the measured pore water pressure minus the hydrostatic pore water pressure. Consequently, as the caisson is lifted, the entire soil plug within the caisson undergoes upward displacement, a phenomenon known as heave. This heave effect, coupled with the suction pressures, significantly enhances the resistance against pullout. Additionally, the surrounding soil is also affected by the suction, mobilizing shear resistance in the vicinity of the caisson. The resistance in this mode is significantly higher than in the frictional mode due to the combined effects of suction, heave, and shear resistance in the surrounding soil. The reverse end bearing failure mode is characterized by a sudden and substantial increase in resistance as the soil plug is lifted as a unit. The third is the intermediate failure mode (partly drained condition) [14]. The partly drained condition represents an intermediate state between the perfectly drained and undrained scenarios. In this mode, the behavior of the caisson during pullout is influenced by both drainage and suction effects to varying degrees. Depending on the degree of drainage, the gap formation between the caisson lid and soil surface may be partial, affecting the extent of skin friction mobilization.

Similarly, the suction pressures within the caisson may not be fully effective in preventing gap formation or inducing heave of the soil plug, resulting in a more reduction in resistance compared to the reverse end bearing mode. The intermediate mode exhibits a combination of features from both the frictional and reverse end bearing modes, with the specific characteristics dependent on the drainage conditions at the time of pullout. Whether drained, partly drained, or undrained behavior occurs certainly depends on the combination of bucket geometry, pullout rate, and permeability of the soil. Due to the slow charging and discharging process of underwater gas storage tanks, the most relevant failure mode is perfectly drained condition. This is also the most dangerous mode.

Overall, many studies have been conducted to investigate the vertical pullout capacity and failure mechanism of regular suction caisson used for traditional marine facilities under different failure modes. However, as an emerging marine structure, little has dealt with how to safely and reliably anchor underwater compressed gas energy storage accumulators to the seabed. In addition, regular suction caisson foundations have some drawbacks, such as the tendency for the bucket to tilt during the installation process, and the formation of local scour pits caused by ocean currents after installation, resulting in a decrease in anchoring force [15,16]. This study presents a novel composite suction caisson used for large-scale underwater compressed gas energy storage to alleviate these problems and studies the pullout capacity of the proposed composite suction caisson foundation under perfectly drained conditions. The pullout capacity is investigated by finite element analysis with ABAQUS/Standard 2020. This paper is organized as follows. In Section 2, the structure composite of an underwater compressed gas energy storage accumulator is introduced. The Finite Element models are established in Section 3. In Section 4, the pullout capacity of regular suction caisson and composite suction caisson is compared and analyzed. Finally, the conclusion is made in Section 5.

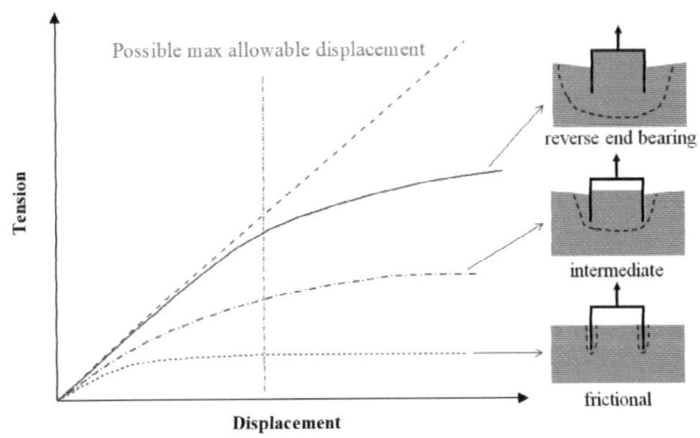

Figure 1. Failure mechanisms according to Senders [9] and Thieken [10]

2. Structural configuration

The structure of the UWCGES accumulator is illustrated in Figure 2. It consists of two main components: a rigid gas storage tank and a composite suction caisson foundation [17]. The top of the gas storage tank is connected to the gas transmission pipeline. The interior of the gas storage tank is equipped with a flexible membrane. This is used for separating seawater from compressed gas to avoid energy loss and contamination caused by the dissolution between gas and seawater. The flexible membrane is typically composed of multiple layers, encompassing an internal matrix material layer and external layers of reinforcement material. Essentially, it should ensure both effective sealing and flexibility. Figure 2 depicts a central pipeline running through the center of the gas storage tank and a gas transportation pipeline connected to it. The central pipeline serves three purposes: maintaining the shape of the flexible membrane, transporting compressed gas, and facilitating the replacement of the damaged flexible membrane. The gas transportation pipeline serves as a link between the gas storage accumulator and the underwater transmission gas pipeline network. There are six ports at the lower position of the storage tank. Seawater can flow into and out of the storage tank via these ports in the release and storage processes. Consequently, the pressure within the large-scale accumulator remains nearly constant, equal to the hydrostatic pressure at the respective water depth.

The composite suction caisson foundation incorporates additional features such as skirts and partition plates, enhancing the design of the regular suction caisson foundation. It consists of an inner and outer bucket, where the inner bucket is partitioned by a plate into three separate chambers. Two holes are arranged on the top of each chamber: one for pressure measurement and the other for water pumping. Deployment begins with the composite suction caisson using its weight to partially penetrate the seabed. When gravity-driven sinking stops, seawater is pumped out of the caisson through designated apertures, creating a vacuum inside. Real-time pressure monitoring is achieved during sinking through pressure gauges positioned in each chamber. Adjustments to the water pumping rate are made based on pressure fluctuations, ensuring controlled sinking in the vertical direction. This gradual, vertical insertion minimizes the risk of post-installation tilting in the seabed soil. Compared to conventional suction caissons, the composite design offers two significant improvements: the outer bucket enhances vertical load-bearing capacity, reduces scouring effects, and prevents downwash erosion. Additionally, the partition plates improve dynamic load-bearing capabilities and enable integrated bucket-soil load distribution.

Figure 3 (a) depicts the energy storage process. Compressed gas and seawater are separated by a flexible membrane. During the storage phase, compressed gas enters the flexible membrane through the gas transmission pipeline, causing the volume of stored compressed gas to increase. Simultaneously, seawater gradually discharges from the storage tank through the seawater ports. Figure 3 (b) illustrates the energy release process. When energy is required, the stored compressed gas is sequentially released through the gas transmission pipeline, the pipeline connected to the outer wall of the gas storage tank, and the underwater gas transmission pipeline network, resulting in a decrease in gas volume. As the gas volume decreases, external seawater flows continuously into the storage tank through the seawater ports.

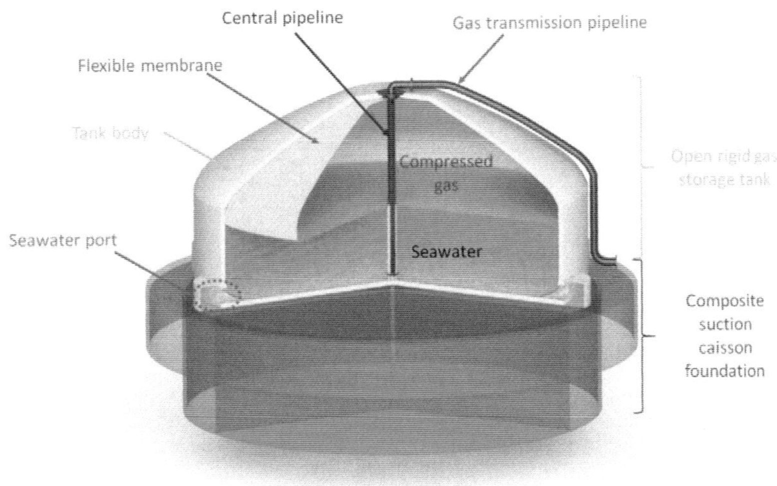

Figure 2. Structural composition of UWCGES accumulator

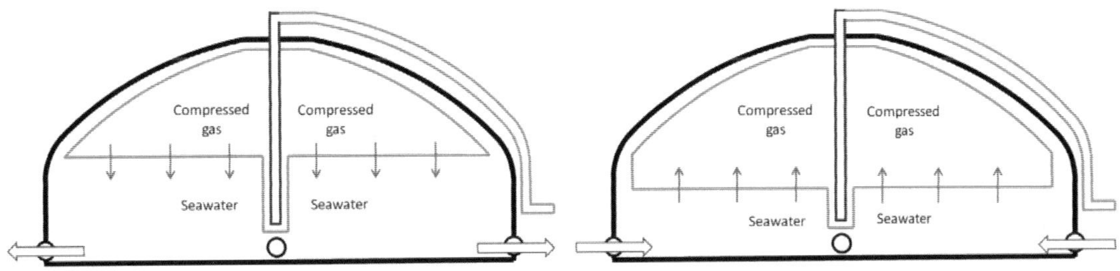

(a) energy storage (b) energy release

Figure 3. The energy storage and release process of UWCGES accumulator

3. Finite Element model

To examine the pullout capacity of suction caisson foundations featuring diverse structural characteristics and aspect ratios, a thorough pullout process analysis is conducted, utilizing the suction caisson depicted in Figure 4. The investigated suction caissons include regular suction caissons and composite suction caissons. The inner bucket diameter is uniformly 10m, and the bucket length is varied to create control groups with aspect ratios of 0.5, 1, and 1.5. The suction caisson has a wall thickness of 0.2 m. The composite suction caisson has a skirt (outer bucket) diameter of 15 m and a height of 3 m. Detailed geometric parameters are presented in Table 1.

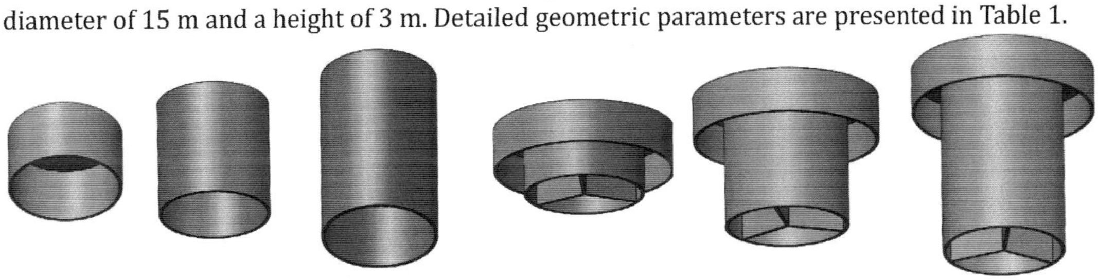

(a) RSC #1 (b) RSC #2 (c) RSC #3 (d) CSC #1 (e) CSC #2 (f) CSC #3

Figure 4. Investigated suction caisson foundation with different length-to-diameter ratios

Table 1. The parameters of the geometric configuration of investigated suction caisson

	Regular suction caisson (RSC)			Composite suction caisson (CSC)		
	RSC #1	RSC #2	RSC #3	CSC #1	CSC #2	CSC #3
Inner bucket diameter (D) [m]	10	10	10	10	10	10
Inner bucket length (L) [m]	5	10	15	5	10	15
Outer bucket diameter [m]	-	-	-	15	15	15
Outer bucket length [m]	-	-	-	3	3	3
Thickness [m]	0.2	0.2	0.2	0.2	0.2	0.2
L/D	0.5	1	1.5	0.5	1	1.5

The finite element models are established in ABAQUS/Standard 2020. Given the symmetry of the suction caisson and soil models, only half of the model is necessary for analysis, as depicted in Figure 5. In this study, the suction caisson foundation is modeled as a concrete structure and assumed to behave as a linear elastic material, neglecting any plastic deformation characteristics. Concrete is one of the most commonly used materials for constructing caisson foundations, especially in civil engineering applications. Concrete offers several advantages such as high compressive strength, durability, and relatively low cost, making it a popular choice for deep foundations, including caissons. Soil is considered as sandy soil material. The elastic-plastic soil is employed following the Mohr-Coulomb failure criterion. The effective friction angle φ is set to 30°. The dilation angle ψ is given as $\psi = \varphi - 30°$. In order to ensure the numerical stability, $\psi = 1°$ is used for $\varphi = 30°$ [12]. The cohesion of sand is reported to be close to zero for saturated sands. Because the use of zero cohesion may lead to numerical instability, typically a small value of cohesion $c = 0.1$ kPa is used in numerical simulations [18,19]. More detailed parameter settings are shown in Table 2.

The boundary conditions in the model are shown in Figure 6. Horizontal displacement constraints are applied to the lateral surfaces of the soil, and displacement constraints in the x, y, and z directions are imposed on the soil's base. Symmetry conditions are enforced across the symmetry plane of the entire model. To accurately establish the initial stress state, apply vertical gravity to the soil and incorporate static water pressure on the soil's surface. A pressure load of 1 MPa is imposed on the soil's surface to mimic the hydrostatic pressure experienced at a depth of 100 meters underwater. The vertical displacement is applied to the foundation. The displacement control method was used because it has been reported to be more accurate in estimating the ultimate bearing capacity compared with a load control method [19,20]. A reference point is established above the foundation surface and coupled to the upper surface. Consequently, vertical displacement is applied directly to this reference point, and the reaction force at the reference point is extracted as the pullout load in Section 4.

The size of the computational domain is chosen to ensure that the boundaries do not exert any influence on the calculated results. The lateral boundary width is set at 7.5D, while the soil depth from the foundation to the bottom boundary is established as 5L. The soil element sizes surrounding the bucket foundation are progressively reduced from the boundary towards the foundation to accurately model the sharp variations in stress and strain near the foundation-soil interface. The soil and suction caisson foundation are both modeled using C3D8 elements. The mesh independence is conducted to ensure accuracy. The tangential mechanical behavior between the soil and foundation surface adheres to the Coulomb friction model, characterized by a friction coefficient of 0.5. The contact formulation between the foundation and the soil is set to surface-to-surface. The contact formulation between the foundation and soil is defined as a surface-to-surface interaction. The foundation surface is designated as the master surface, while the soil surface serves as the secondary surface. The interface element between the suction caisson and the soil is defined comprehensively to capture both the normal and tangential behavior accurately. For the normal behavior, a "hard contact" formulation is employed. Regarding the tangential behavior, the penalty method is adopted. The detailed contact settings can be found in Table 3.

The FE analysis procedure comprised three distinct phases. Initially, geostatic stresses were imparted to the unconstrained soil elements. Subsequently, the soil components within the bucket's configuration were substituted with foundation elements, activating the interface in the process. Notably, the installation sequence of the bucket foundation was not modeled, with the foundation considered as "wished-in-place." While soil attributes such as density and effective stress might undergo modifications during installation, particularly proximate to the foundation, their impact on the derived bearing capacity was presumed to be negligible. Finally, the foundation's uppermost displacement was incrementally increased until the structure reached its failure point [19].

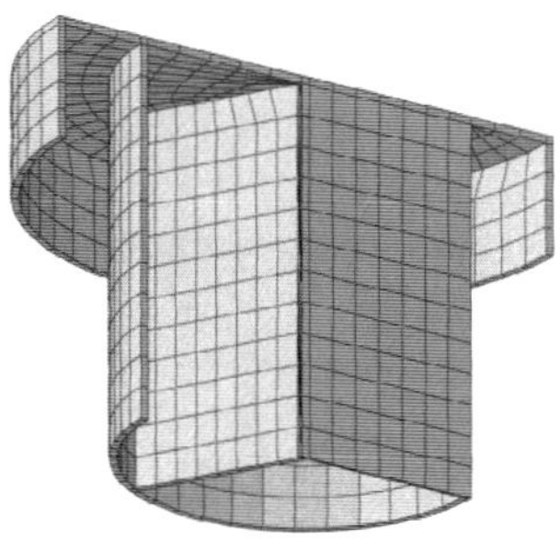

(a) composite suction caisson foundation

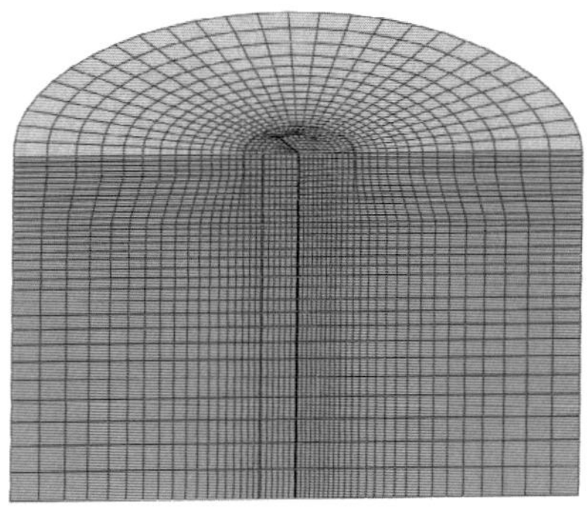

(b) soil

Figure 5. 1/2 model in ABAQUS/Standard 2020

Table 2. The parameters of materials

	Density [kg/m^3]	Young's modulus [Pa]	Poisson's ratio	Cohesion [Pa]	Frictional coefficient	Effective friction angle φ [°]	Dilatancy angle [°]
Suction caisson	2400	25e9	0.2	-	0.5	-	-
Soil	1800	60e6	0.3	100	0.5	30	1

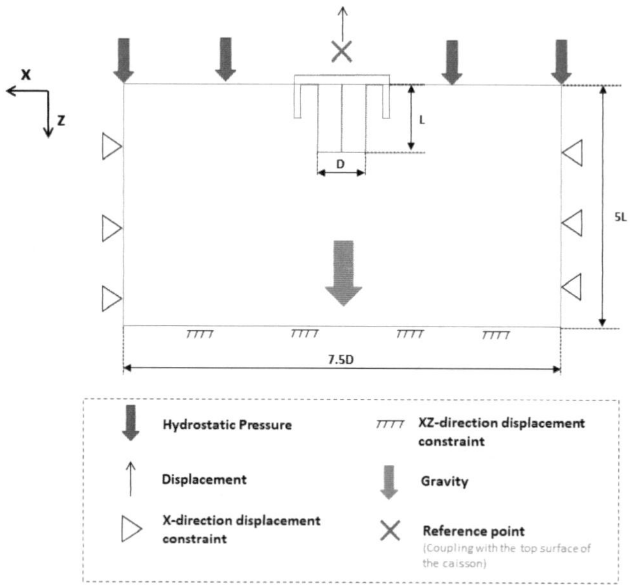

Figure 6. Boundary condition

Table 3. Detailed mesh and interaction properties

<table>
<tr><th></th><th>Element Type</th><th>Plasticity</th><th>Bucket–Soil interface</th><th colspan="2">Contact formulation</th></tr>
<tr><td>Suction caisson</td><td rowspan="2">C3D8</td><td>-</td><td rowspan="2">Coulomb friction</td><td rowspan="2">Surface-to-surface</td><td>Main surface</td></tr>
<tr><td>Soil</td><td>Mohr-Coulomb</td><td>Secondary surface</td></tr>
</table>

4. Results and Discussion

4.1 Geostatic equilibrium

Ground stress is the natural stress that exists in the soil without disturbance. Achieving geostatic equilibrium involves attaining a state characterized by the presence of initial stress without accompanying initial strain. To obtain a more precise initial stress state, it is crucial to balance both the soil's gravity field and the initial stress field. The challenges associated with achieving accurate geostatic equilibrium, particularly in complex scenarios involving significant interactions between the soil and the engineering structure are recognized. Therefore, a two-step approach, leveraging the model change functionality in ABAQUS, in conjunction with the automatic equilibrium method is adopted. This approach aims to improve both the convergence and accuracy of the initial stress state. In the first step, the soil domain is isolated by the deactivation of the caisson model, and an automatic geostatic equilibrium analysis is conducted solely on the soil. This step accurately captured the initial stress state of the soil prior to the installation of the caisson, taking into account the effects of gravity and boundary conditions. Subsequently, in the second step, the caisson model is reinstated by its activation, while simultaneously, the soil elements within its footprint are deactivated. This setup facilitated the definition of realistic contact conditions between the caisson and soil, reflecting the post-installation scenario. Another equilibrium analysis is then performed, incorporating the interactions between the caisson and soil, to simulate the stable state of the system after the installation of the caisson.

During simulation calculations, the deformation of the soil due to its self-weight and the initial stress is minimized to ensure minimal impact on subsequent calculations. In this study, two primary initial stresses require balancing: the self-weight stress arising from the soil's gravity, and the hydrostatic pressure induced by the depth of the surrounding seawater. This approach ensures an accurate representation of the initial stress state in the simulation. The criterion for assessing the adequacy of geostatic equilibrium is set as follows: if the soil's displacement remains within the order of 10^{-4} meters, indicating minimal deformation, the achieved geostatic equilibrium is deemed acceptable. This threshold ensures that the simulation accurately reflects the undeformed, initial stress state. The results of the geostatic equilibrium analysis are presented in Figure 7. Specifically, Figure 7(a) depicts the initial displacement field, while Figure 7(b) illustrates the stress distribution in the soil after geostatic equilibrium. Analyzing Figure 7(a), it is evident that the maximum displacement is on the order of 10^{-15} meters, significantly smaller than the predetermined threshold of 10^{-4} meters. It indicates that the geostatic equilibrium achieved is exceptionally good, thereby validating the quality of the address quality geostatic equilibrium effect.

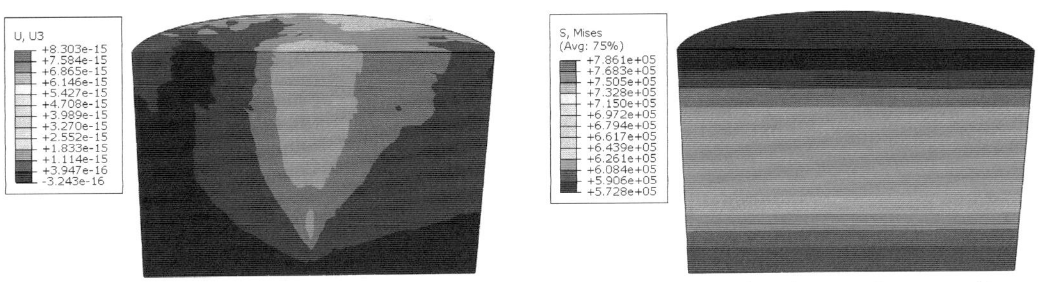

(a) Initial deformation

(b) Stress distribution after geostatic equilibrium

Figure 7. Contour Plot of Geostatic Equilibrium for CSC #2

4.2 Pullout process

Figure 8 depicts the displacement contour plot of the entire model during the pullout process of CSC # 2. The displacement distribution of the model at different positions can be obtained through this figure. Observing color changes in the cloud map enables intuitive identification of areas with strong and weak displacement in the model, along with discernible trends and patterns in displacement changes. Figure 8 depicts the suction caisson foundation with a red coloration, signifying the maximum displacement. A gradual transition from red and yellow to blue-green occurs radially from the center, illustrating the decreasing trend of displacement with increasing distance. Specifically, the soil displacement near the caisson's top is greater, tapering off gradually as the distance from the caisson increases.

The foundation is pulled out, driving the soil to move upwards. The maximum displacement of the foundation is 0.25m, and the soil undergoes plastic failure. The contours of foundation displacement relative to the soil and soil flow mechanisms are shown in Figure 9. The colors and arrows indicate the magnitude and direction of soil displacement, respectively. The more pronounced colors typically signify greater displacement. The figure highlights that the soil closest to the foundation or caisson experiences the most significant displacement. This

Journal of Physics: Conference Series **2929** (2024) 012010 doi:10.1088/1742-6596/2929/1/012010

is due to the direct interaction between the soil and the caisson's surface as it is pulled out. The soil immediately surrounding the caisson is visibly disturbed, with an upward and outward flow pattern as the caisson is lifted. As the caisson is extracted, the soil movement propagates outwards. The soil farther away from the caisson is less directly affected but still experiences some upward movement, primarily driven by the movement of the soil immediately adjacent to the caisson. This demonstrates the interconnectedness of soil layers and how disturbances can propagate through the soil mass.

Soil, as an elastic-plastic material, undergoes deformation in response to applied external forces. When deformation exceeds a critical threshold, the soil undergoes plastic yielding, potentially failing. Figure 10 depicts the yielding behavior of soil during foundation extraction. Utilize distinct colors in the figure to delineate regions that have attained the yield state. The yielding process of soil during the process of foundation extraction gradually develops from local to global, from shallow to deep. In the early stage of foundation extraction, the soil inside the bucket in contact with the foundation first yields. In the middle stage, the soil around the top surface of the foundation and the soil at the end of the caisson yield. In the later stage, the yield zone gradually expands. The yielding process reflects the process of soil resistance exerted, with skin friction exerting its effect first and end resistance exerting its effect later. This is a significant feature of the composite suction caisson foundation.

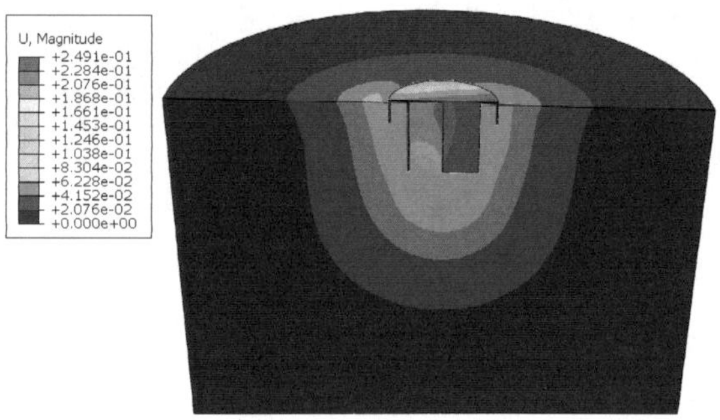

Figure 8. Displacement contour plot of CSC #2

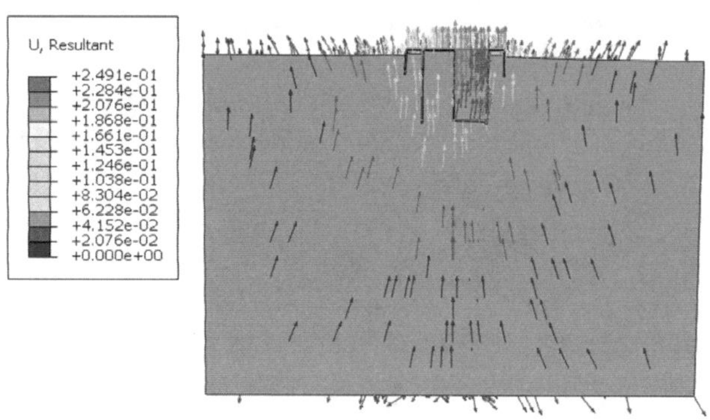

Figure 9. Contours of foundation displacement relative to the soil and soil flow mechanisms

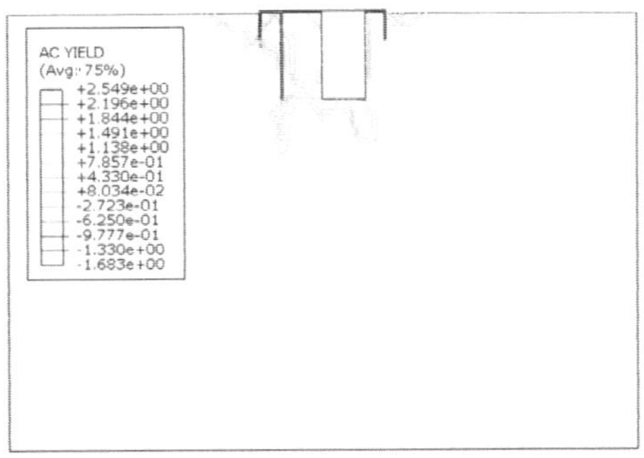

Figure 10. Soil yield zone of CSC #2

4.3 Pullout capacity

The pullout capacity is estimated by applying the tangent intersection method illustrated in Figure 11 [21-23]. Two tangents are drawn to touch the early and later sections of the curve. The coordinate on the curve corresponding to the intersection of the tangents defined the pullout capacity and the corresponding displacement. In Figure 11, the vertical axis depicts the pull-out load, whereas the horizontal axis indicates the displacement of the caisson. To enable a straightforward comparison among various foundation types, the pullout displacement was normalized. The horizontal axis is denoted as h/D, where 'h' represents the displacement of the caisson, and 'D' corresponds to the diameter of the inner bucket. As the displacement of the caisson augments in Figure 11, the pullout load undergoes a gradual increase. Employing the tangent intersection method, the ultimate pullout load is estimated to be 7E+07N, and the location where the ultimate pull-out occurs is at h/D=0.016.

Figure 12 illustrates the correlation between the pull-out load and displacement of the investigated suction caisson. An augmentation in the length-to-diameter ratio (L/D) of the caisson corresponds to an elevation in the ultimate pullout load of the foundation. The primary contributor to the pullout capacity of the foundation under drainage conditions is the frictional force between the soil and the caisson wall. An increase in the length-to-diameter ratio results in an expansion of the contact area between the soil and the caisson, subsequently enhancing the pullout load. Furthermore, regular suction caisson foundations with an equivalent aspect ratio exhibit a lower ultimate pullout load compared to composite suction caisson foundations, attributable to the latter's larger contact surface.

The ultimate pullout capacity and corresponding displacement of the investigated caisson are shown in Table 3. Adding outer bucket and partition plates based on regular suction caissons can significantly improve the pullout capacity. The pullout capabilities of CSC #1, CSC #2, and CSC #3 increased by 100%, 29.6%, and 12.8%, respectively, compared to their corresponding RSC counterparts (RSC #1, RSC #2, and RSC #3). As L/D increases, the trend of increasing pullout capacity decreases. Similarly, the normalized displacement of the caisson with ultimate pull-out capability also increases with the increase of the length-to-diameter ratio, especially with the proposed CSC. In comparison to RSC, CSC with varying aspect ratios show a more pronounced divergence in normalized displacement related to their ultimate pullout capacity. Nonetheless, the variation of displacement for the caisson at its ultimate pullout capability is relatively small, between 0.01D and 0.027D.

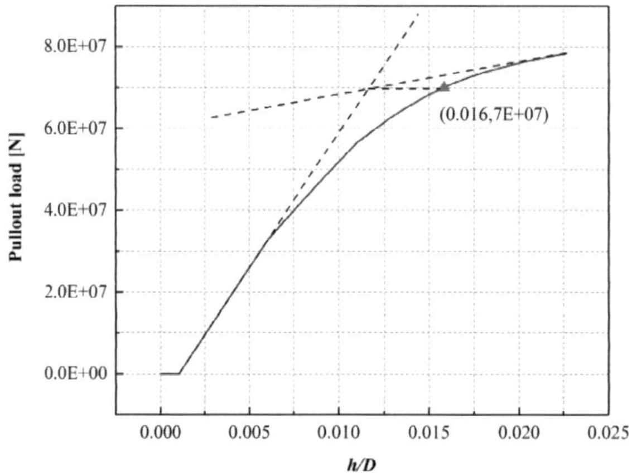

Figure 11. Estimation of the pullout load of CSC #2

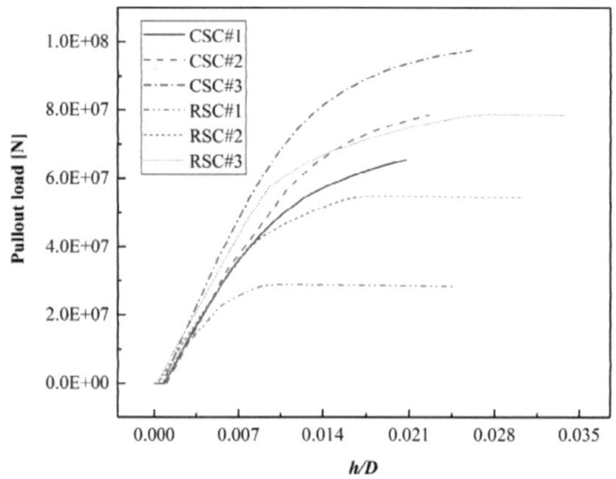

Figure 12. The pullout load of the investigated suction caisson foundation

Table 4. Ultimate pullout capacity and its corresponding caisson displacement

Suction caisson	Ultimate pullout load [N]	h/D
RSC #1	2.9E+07	0.01
RSC #2	5.4E+07	0.016
RSC #3	7.8E+07	0.027
CSC #1	5.8E+07	0.014
CSC #2	7E+07	0.016
CSC #3	8.8E+07	0.017

5. Conclusions

In this study, a composite suction caisson foundation used for large-scale underwater compressed gas energy storage accumulator is proposed, and the pullout capacity and pullout process analysis are conducted. The major findings of the study are summarized as follows.

- Frictional Force Dominance: The frictional force between the bucket wall and the soil is the main factor contributing to the pullout capacity. During the pullout process, skin friction exerts its effect first, and the end resistance exerts its effect later.
- Foundation Height Impact: The pullout load and the corresponding displacement of the caisson both increase with the foundation height (L). Notably, the displacement at ultimate pullout capacity remains relatively small, ranging from 0.01D to 0.027D.
- Enhanced Performance of CSC: The proposed CSC, featuring an outer bucket and partitions, effectively mitigates issues such as bucket tilting during installation and local erosion during operation. Compared to the RSC, the CSC exhibits a significant improvement in pullout capacity, with an increase ranging from 12.8% to 100% for varying L/D ratios. However, as L/D increases, the trend of enhanced pullout load diminishes.

Overall, the proposed CSC exhibits excellent pullout capacity compared to RSC, primarily due to its structural improvements. These improvements effectively alleviate issues such as bucket tilting and local erosion commonly encountered in engineering applications, while also enhancing the overall pullout capacity. Nevertheless, it should be noted that this study focuses solely on the third mode of pullout failure, overlooking potential complexities and alternative failure modes in real-world applications. When the pullout speed is too fast, the suction pressure generated under the top cover of the suction caisson foundation accounts for more than 50% of the pullout capacity. The influence of pullout speed and suction pressure on pullout capacity should be conducted. Additionally, this study is preliminary and relies solely on simulation, necessitating experimental validation for confirmation.

References

[1] Wu X, Hu Y, Li Y, Yang J, Duan L, Wang T, Adcock T, Jiang Z, Gao Z, Lin Z, Borthwick A and Liao, S 2019 *RENEW SUST ENERG REV* 104 379-393
[2] Stapelfeldt M, Bienen B and Grabe J 2020 *Ocean Eng* 215 107105
[3] Zhao J, Li D, Zhang Y, Wu Y and Qiu P 2024 *Ocean Eng* 309 118602
[4] Malik B A, Mandhaniya P, Soni A K and Ouyang Z 2024 *Ocean Eng* 309 118491
[5] Seo S, Mun H, Lee J and Kim J 2015 *MAR STRUCT* 44 142-158
[6] Chamelia D M, Wardhana W and Prastianto R W 2015 *PROG EARTH PLANET SC* 14 220-227
[7] Wang Z, Wang H, Sant T, Zhao Z, Carriveau R, Ting DSK, Li P, Zhang T and Xiong W 2024 *Int J Hydrogen Energy* 71 1266-1282
[8] Wang H, Wang Z, Liang C, Carriveau R, Ting DSK, Li P, Cen H and Xiong W 2022 *Appl Sci* 12 9361
[9] Senders M 2008 Ph.D. thesis The University of Western Australia, Australia
[10] Thieken K, Achmus M and Schröder C 2014 *Comput Geotech* 60 88-100
[11] Deng W, Carter, J 2002 *Int. Offshore Polar Eng* 12 89–97
[12] Wang X, Yu L, Yang Q and Tang X 2022 *Ocean Eng* 266 113169
[13] Li D, Ma S and Zhang Y 2019 *MAR GEORESOUR GEOTEC* 37 291-300
[14] Sun L, Qi Y, Feng X and Liu Z 2020 *Ocean Eng* 197 106893
[15] Yao W, Draper S, An H, Cheng L, Harris J M and Whitehouse R J S 2020 *Ocean Eng* 218 108127
[16] Wei K, Qiu F and Qin S 2022 *Ocean Eng* 250 111052
[17] Wang H, Xiong W, Liang C, Carriveau R, Ting D S K and Wang Z 2023 7th Offshore Energy & Storage Symposium (OSES 2023) 15-23
[18] Wang X, Yang Q, Tang X 2022 *MAR GEORESOUR GEOTEC* 40 36-51
[19] Park J S, Park D and Yoo J K 2016 *Ocean Eng* 121 453-461
[20] Gourvenec S, Randolph M 2003 Geotechnique 53 575–586
[21] Vicent S, Kim S R 2020 *Appl Ocean Res* 101 102198
[22] Ssenyondo V, Hong S, Bong T and Kim S R 2021 *Ocean Eng* 237 109643

Journal of Physics: Conference Series **2929** (2024) 012010 doi:10.1088/1742-6596/2929/1/012010

[23] Zhu B, Dai J, Kong D, Feng L, Chen and YM, M Y 2019 *Can Geotech J* 1293–1303

8th Offshore Energy & Storage Symposium (OSES2024) IOP Publishing
Journal of Physics: Conference Series **2929** (2024) 012011 doi:10.1088/1742-6596/2929/1/012011

Grand Cayman's Goal of 70% Sustainable Electricity: A Case Study

James Ebanks*[1], and Daniel Friedrich[2]

[1]Department of Engineering, Caribbean Utilities Company, George Town, Grand Cayman
[2]Chair of Energy Systems, University of Edinburgh, Edinburgh, Scotland

E-mail: jaebanks@cuc.ky

E-mail: D.Friedrich@ed.ac.uk

Abstract. Renewable energy sources are a key technology for reaching net zero carbon emissions and reducing the contribution to climate change, but there are challenges and consequences of integrating large amounts into a grid. Grand Cayman is a Caribbean island with abundant solar resources and an ambitious energy policy, but neither the effect of the grid nor the implementation of the energy policy has been publicly modelled and analysed as of yet. In this research, the grid was modelled, and two solutions were designed and compared. The first solution was composed of mainly solar PV and aimed to follow the national energy policy while the second attempted to offer a stable base-load by using OTEC (Ocean Thermal Energy Conversion). The goal was to increase the share of DG (Distrusted Generation) to 70% of the island's total electricity while not decreasing power quality. Both resulted in an improved voltage magnitude, power losses, and branch limits compared to the existing grid. This research provides a groundwork for further investigation to examine minute fluctuations throughout the day or seasonal changes throughout the year.

1 Introduction

1.1 The Impact of Climate Change

The Cayman Islands is one of the many islands in the Caribbean that depend on its tourism industry which is severely impacted by climate change. The effects of climate change, such as rising sea levels, increasingly dangerous storms, and degrading coral reefs, have become more apparent through the decades. These effects significantly impact the Caribbean on an ecological and economic level [1].

In addition to environmental improvements such as reducing local emissions of fine particular matter (PM2.5) and other pollutants from diesel, other benefits of switching to renewable sources include potentially cheaper electricity and less dependence on international oil markets. In 2021, the average price of electricity in the United States was 13.72 cents per kWh, compared to Grand Cayman's average price of 29.4 cents per kWh [2], [3]. Due to Russia's invasion of Ukraine, global oil prices have increased drastically and the average price of electricity in Grand Cayman has increased from 29.4 cents per kWh in 2021 to 34.5 cents per kWh in July 2022 mainly due to increased fuel costs [4], [5]. Decreasing electricity prices are seen as a significant benefit to the economy and have been mentioned across policies for small island nations such as the Bahamas and the Cayman Islands [6], [7].

1.2 General Overview of the Cayman Islands
The Cayman Islands comprises of three islands called Grand Cayman, Cayman Brac, and Little Cayman. They are located in the Caribbean, south of Cuba, and west of Jamaica, with a population of 71,432 as of 2021 [8].

The 2017 National Energy Policy (NEP) of the Cayman Islands represent the fundamental key goals for sustainable and efficient energy production and consumption of the Cayman Islands from 2017 to 2037 [7]. The main goal for the electricity sector is to supply 70% of all electricity generation from renewable energy sources by 2037. Furthermore, the goal states that the reliability of the national grid must not be compromised when integrating renewable generation by investing further into system stability. As of 2015, Renewable Energy (RE) accounted for 0.9% of electricity generation, and 99.1% is from diesel generation. The projected yearly electricity generation for 2037 (from 2015) can be seen in *Table 1.*

Table 1: Forecast Potential Sources of Annual Electricity Generation in 2037 (forecasted from 2015) [7]

TYPE OF GENERATION	GENERATION (MWH)	(%) OF TOTAL GENERATION
Diesel Generators	216,616	30
Wind Power	24,528	3
PV - Utility Scale	450,292	62
Waste-To-Energy	21,900	3
PV - Distributed	17,086	2
Total	**730,423**	**100**

1.3 Challenges of Integrating RE
The most well-known issue of high Variable Renewable Energy (VRE) generation penetration is the wind and solar resource variability. The power output of VRE generation plants can widely vary in seconds due to the changing weather conditions such as sunlight and wind speed. This variability has adverse effects on both the voltage and frequency by causing rapid fluctuations. On the other hand, conventional synchronous generation can control the frequency and voltage fluctuations due to inherent inertia and reactive power control. The sizeable rotating machinery can absorb or release kinetic energy to balance the electrical and mechanical power of the generator [9]. Thus replacing more conventional generations requires supplementary power quality support. General solutions for rapid voltage and frequency fluctuations include the installation of high discharging energy storage such as lithium-ion batteries. [10].

In addition, the duck curve is one of the highest profile issues with the large penetration of PV systems. This occurs when the maximum energy output by PV systems does not occur during the most demanding times of the day. This disparity leads to a mismatch in load and demand. Solutions such as supplementary generation and energy storage are required to meet demand during the night [11]. These challenges show that it is essential to carefully evaluate and design the transition to VRE systems so that it provides environmental and commercial benefits.

1.4 Research Goal
This research aims to answer the central question: What are some of the challenges when integrating a high amount of Distributed Generation (DG) in Grand Cayman? Furthermore, it shows how these challenges can be addressed by conventional RE and by incorporating other viable new technology such as Ocean Thermal Energy Conversion (OTEC). The scope of the research was limited to an hourly time-scale and only focused on the transmission system. The main objectives were to compare the two solutions based on their respective voltage quality, branch limits, power losses, storage solutions, and other integration requirements.

2 Material and Methods
This section covers the modelled system, methodology, and specific methods that were used to simulate the grid, analyse various cases, and measure the impact of DG integration. This methodology can be applied to other islands.

2.1 Tools

MATPower is a free, easy-to-use, and open-source tool for Power Flow Optimization that has standard iterative power flow methods and flexibility. It required a generation, branch, and load matrix which modelled a given grid [12]. Newton-Raphson method was used to solve the power flow as it is more accurate when computing large grids.

2.2 Key Measurement Indices

Standard indices were used to evaluate DG's impact on the system's active power loss, reactive power loss, and voltage magnitude [13].

- Loss Sensitivity Factors (LSF) measure the total active and reactive power loss in a system before and after a DG has been placed. *Eq. 1* and *Eq. 2* display the active power (PLF) and reactive power loss factor (QLF), respectively, where N is the number of buses in the system [14].

$$PLF = \frac{\sum_{i=1}^{N} (P_{Lineloss}(i))_{after\ DG}}{\sum_{i=1}^{N} (P_{Lineloss}(i))_{before\ DG}} \tag{1}$$

$$QLF = \frac{\sum_{i=1}^{N} (Q_{Loss}(i))_{after\ DG}}{\sum_{i=1}^{N} (Q_{Loss}(i))_{before\ DG}} \tag{2}$$

- Voltage Improvement (VI): VI measures the voltage difference between bus i and its voltage setpoint before and after DG has been installed and is expressed in *Eq. 3*, where Vi,ref is the reference voltage of bus i.

$$VI = \frac{\sum_{i=1}^{N} |V_i - V_{i,ref}|_{afterDG}}{\sum_{i=1}^{N} |V_i - V_{i,ref}|_{beforeDG}} \tag{3}$$

- Voltage Profile: This general index measures the voltage difference at bus i and its voltage setpoint, as shown in *Eq. 4*.

$$V_{pro} = \sum_{i=1}^{N} (V_{pro} - V_{rated})^2 \tag{4}$$

- Voltage Deviation (VD): ΔV_d measures the voltage deviation between the reference bus and chosen bus. This factor can help quantify the effects on voltage quality due to long transmission connections and heavy loads and is described below in *Eq. 5*. Where $V_{i,ref}$ is the voltage of the reference bus.

$$\Delta V_d = \sum_{i=1}^{N} \left(\frac{V_{i,ref} - V_i}{V_{1,ref}}\right)^2 \tag{5}$$

Other indices such as economic, environmental, and distribution network voltage quality were not considered due to the scope of research and lack of data.

2.3 Algorithms

Bawazir and Cetin completed a comprehensive review of 13 algorithms that aim to optimize PV-DG placement in a power grid [13]. These algorithms used a mixture of indices that measured emissions, costs, the environment, and power quality (mentioned above). Bawazir and Cetin concluded that real power loss and voltage stability index have higher importance compared to the others. They compared complex optimization algorithms and found the amount of time and computational power needed to provide a solution can exponentially increase depending on how large the grid model is and how many PV-DG installations are required [13]. An algorithm which was quicker for computation and easier to execute was needed as the Grand Cayman model system had over 100 buses and required 4 types of DG in the solutions.

2.4 Greedy Strategy Algorithm

The Greedy Strategy Algorithm (GSA) is an iterative algorithm that determines the locally optimal solution at each iteration to build a global solution. Although this algorithm is fast and easy to implement, the combination of locally optimal solutions does not guarantee a globally optimal solution. Furthermore, GSA does not find the optimal size of DG, instead, a specific amount of DG needs to be selected, and then applied to the system to measure its impact.

2.5 The Modelled System

Each island has its independent grid and generation as none of the islands are interconnected. The main focus of this research was conducted on Grand Cayman, the largest of the three. The national grid in Grand Cayman is owned and operated by the largest generation provider, Caribbean Utilities Company (CUC) which provided all the relevant grid data required for the model. As of 2020, the system can be described as a two-loop system encompassing the west and east end of the island. The modelled system contains 18 feeders, 7 substations, 83.15km of transmission cables, 24 generators, 88 buses, and 98 branches. All generators are concentrated in a central plant in the middle of the island. The grid has two voltage levels for transmission systems (69kV and 13kV) and many voltage levels for the distribution systems [3].

The bus with the highest power output was chosen as the reference bus (bus 96). To accurately model the grid, assumptions were made, such as defining MVA branch limits based on conductor gauges. The data was then adapted to fit within MATPower's required matrices. It was decided that the most relevant buses for modelling were the 69kV, 13kV and 5kV as buses with lower voltage either had an insignificant auxiliary load or were in the distribution network which was out of the research scope. The modelled system without any addition of DG was called the base case scenario.

2.6 Research Methodology

2.6.1 Developing Solutions To develop a solution a load profile detailing the load at each feeder was required. CUC provided load data from 2016 to 2020, and it was determined that the most suitable development environment was during peak solar irradiance (12:00) because the NEP requirements had a large amount of solar. Examining the data provided by CUC, the largest load that occurred at 12:00 was on August 28th, 2019 measuring a total of 101.9 MW and 37.6 MVAr. Using the load profile guaranteed a large amount of energy transfer as solar supply was at the peak and load was near the peak.

2.6.2 Designing the GSA A simplified flow chart of the GSA can be seen in *Fig. 1*. The GSA was designed to take two main sets of inputs which were the MATPower matrix requirements and the following DG information:

- Type: The type of DG can include OTEC, Waste to Energy, Wind, or PV, (installed in this order) which allowed for a mix and match approach and a flexible solution.
- Increment MW: The amount of additional DG added at every bus. After a successful iteration, this Increment MW is added to the optimal bus permanently.
- Max MW: The total maximum allowable MW for each type of DG
- Power Factor: The desired power factor for each type of DG.

Once the inputs sets were defined and accepted, one at a time, each type of DG was injected into the grid. The algorithm started with OTEC generation, then waste-to-energy as these are generally constant and not modular, then moved to PV and Wind. This installation sequence allowed for any voltage disturbances or power loss increases due to the implementation of a centralized OTEC or waste-to-energy plants to be addressed by the more modular PV and Wind systems as they allowed for multiple array sizes to be installed on multiple buses. The amount of energy from OTEC was relatively arbitrary and chosen to be 15% of the total energy supply. The choice of 15% resulted in an Increment MW and Max MW of 14.23MW (rounded to 15MW) as it was assumed the OTEC would constantly be running for 24 hours a day. For the waste-to-energy system, there were recent agreements for a single plant with a power capacity of 8.5MW (rounded to 9MW) [15].

Every bus was injected with an increase of Increment MW one at a time, and then the power flow would be solved in the development environment discussed in the section above. Afterwards, the impact indices (PLF, QLF, VD, VI, and Vpro) were measured, calculated, and scored. Each index had a different weight based on importance, and this can be seen in *Table 2*.

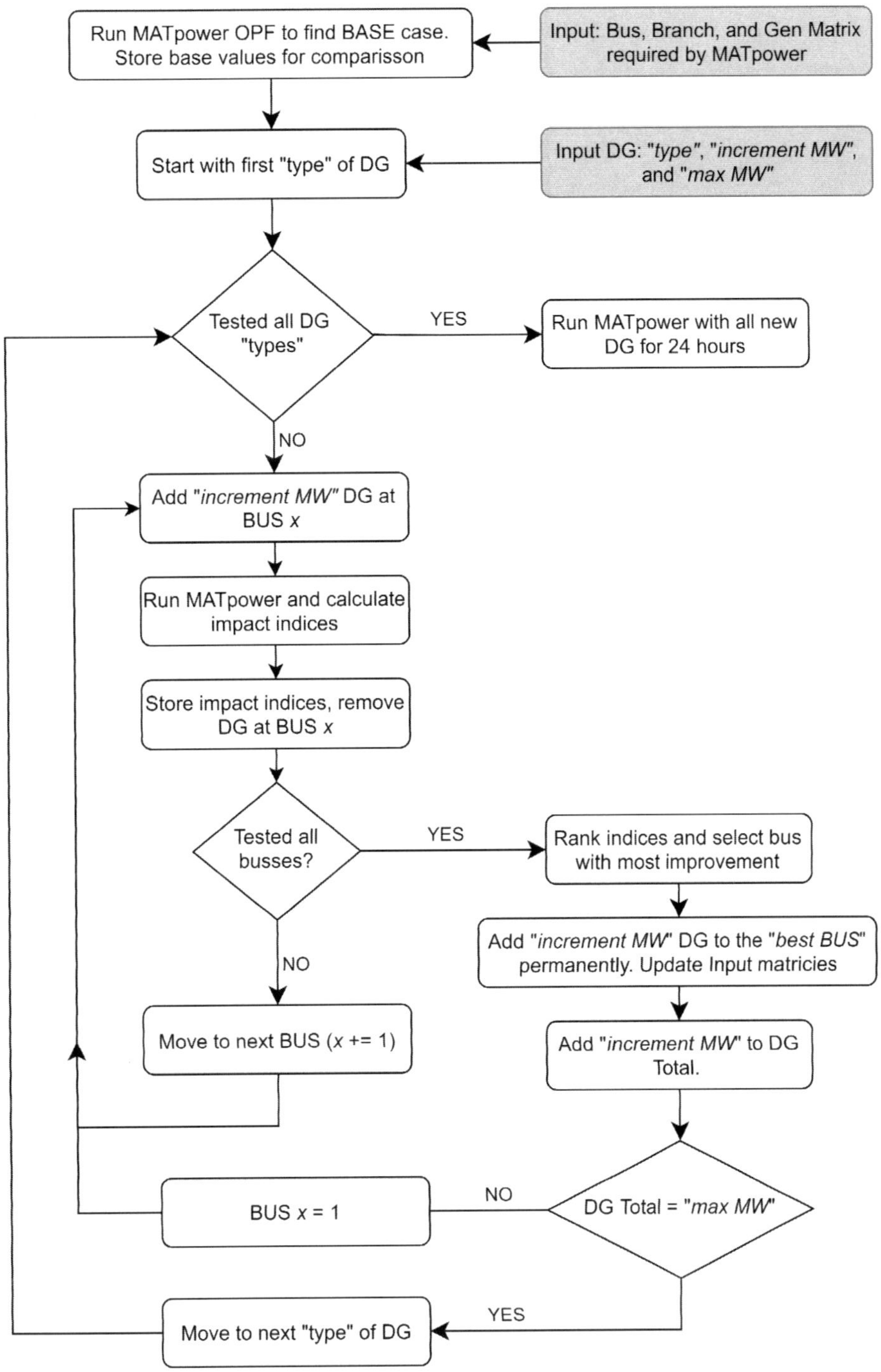

Figure 1: Basic flow diagram of the Greedy Strategy Algorithm

Table 2: Impact indices and their corresponding weights

IMPACT INDEX	WEIGHT
Active Power Loss Factor (PLF)	0.4
Reactive Power Loss Factor (QLF)	0.1
Voltage Deviation (VD)	0.2
Voltage Improvement (VI)	0.2
Voltage Profile (Vpro)	0.1

The weight of the indices was from, Bawazir and Cetin's comprehensive review [13]. They determined Active power loss was the most critical index as these losses manifested as heat in the branches, which caused a decrease in efficiency. The indices were multiplied by their respective weights and summed together to form a score, seen in *Eq. 6.* Where i is the bus, and σ is the weight for each index.

$$\begin{aligned} SCORE_i = PLF_i * \sigma_{PLF} + QLF_i * \sigma_{QLF} \\ + VD_i * \sigma_{VD} + VI_i * \sigma_{VI} \\ + Vpro_i * \sigma_{Vpro} \end{aligned} \quad (6)$$

After each bus has been injected and the indices scored, the best bus was determined as the one with the lowest score, i.e. the highest positive impact on the system. For instance, a PLF of 1 indicated no change, while a PLF of 0.7 indicated the active power losses were reduced by 30% compared to the base case. The lower the indices, the more improvement in the system compared to the base case.

Once the best bus was chosen, it was permanently added as part of the system. The system would iterate until the Max MW was reached for that DG type, then move to the next Type with its corresponding Increment MW, Max MW, and Power Factor. As more DG is added into the system, conventional generation is reduced unit by unit. The units with the most power losses are the first to be turned off.

2.6.3 Testing Method The base case is the current transmission grid for Grand Cayman which contained a maximum generation capacity of 183.23 MW across 24 conventional generation systems. A load profile was then created to measure the performance of the base case or any developed solution. The load data provided by CUC was examined, and every hour of every day was analysed to determine a minimum, average, and maximum hourly load. The test environment, to compare all cases, was chosen to be the peak day. It was defined as the day with the most energy usage, which also corresponded with the highest peak in CUC's history, 113.47MW and 41.86 MVAr at 17:00 on August 28th, 2019.

Once all types of DG's have been installed to the desired Max MW, a peak day simulation was conducted. The optimal solution was tested for 24 hours during the peak day. Each hour, the storage systems, and conventional generation fluctuated during the day to meet the fluctuating load. DG systems were programmed on a time basis and outputted at their maximum, conventional generation and OTEC ran 24 hours a day at a fixed output, and the storage systems produced when generation was lacking or absorbed the excess generation. It was assumed that PV installations were DC coupled with the energy storage to allow both systems to use the same inverter installation. Each storage system matched the power output of the PV systems, i.e both a 2MW PV system and a 2MW storage system be would installed at the same bus. The energy capacity, efficiency, and type of storage were not chosen as detailed storage solutions were out of scope. As expected during the GSA process, most of the conventional generation was turned off and disconnected as they were no longer needed.

3 Results

In this section, two solutions are discussed and compared against the base case. The first is a NEP solution that follows the specific energy portfolio outline in Section 1.2. The second is a custom solution using OTEC as a baseload. Their voltage profile, power losses, branch limits, and storage requirements are compared.

3.1 NEP Solution
A simplified generation schedule was made to determine the Max MW per DG. This schedule was designed for NEP's specified energy portfolio and was simulated on the peak day scenario. The Max MW for solar, wind, and waste-to-energy, are 193.8MW, 10.17MW, and 8.54MW, respectively. Max MW per DG was rounded up to the nearest whole number, and the Increment MW was set based on the type. Wind and solar had an increment of 1MW as they are more modular and can be installed on the bus at multiple iterations; however, waste-to-energy is predefined at one plant at 9MW. The input to the algorithm can be seen in *Table 3* below.

Table 3: GSA DG input parameters for the NEP solution

TYPE	OTEC	PV	WIND	WtE
Increment MW	0	1	1	9
Max MW	0	194	11	9
PF	0	0.95	0.8	0.8

The total installed capacity of DG was 214MW. Solar accounted for 194MW, wind for 11MW, and waste-to-energy for 9MW. As expected, most of the DG was placed nearest to the loads to reduce power losses in transmission lines and voltage fluctuations. The percentage of DG and load at each bus were closely matched. For instance, bus 71 had the highest percentage of DG and load, corresponding to 21.36% and 20.18%, respectively. In total, 99.07% of the total installed DG was located on a bus with a load. Appendix Table 8 contains a breakdown of DG installations at each bus.

3.2 OTEC Solution
The OTEC solution was similar to the NEP solution; however, the biggest difference was a 15MW OTEC plant was used to reduce storage system capacity and offer a base load.

OTEC uses the difference in temperature of the ocean surface and the deep layers to produce electricity. The warmer surface water vaporizes the working fluid (typically ammonia), which is used to drive the turbine [16]. The cold water is then extracted from the deep layers and used to re-condense the working fluid. According to a brief published by IRENA(International Renewable Energy Agency) the Cayman Islands is located in a high potential area as the temperature difference between the normal depths used for OTEC is high. Due to this high difference OTEC can provide a large baseload for the electricity demand. Although the largest currently operating OTEC plant is only 100 kW there are many other plants in various phases of development across the globe [17].

Similar to the NEP solution, a generation schedule was determined and the Max DG were then rounded up to the nearest whole number. As OTEC and waste-to-energy were considered to have one plant each, their Increment MW was equal to their Max MW. Similar to the NEP solution, the waste-to-energy and wind remained the same. However, the solar's Max MW was decreased from 194 MW to 149 MW because the OTEC is replacing a portion of the PV to provide a stronger base load. The input to the GSA can be seen below in *Table 4*. Appendix Table 7 contains a breakdown of DG installations at each bus.

Table 4: GSA DG input parameters for the OTEC solution

TYPE	OTEC	PV	WIND	WtE
Increment MW	15	1	1	9
Max MW	15	149	11	9
PF	0.8	0.95	0.8	0.8

3.3 Comparison of NEP Solution, OTEC Solution, and Base Case
The section aims to compare the results of the implemented NEP solution, OTEC solution, and the base case. All three systems were simulated in the same test environment described in Section 2.6.3. The

power losses, voltage magnitude, and branch limits are compared. In short, the voltage and power losses of the system have drastically improved compared to the base case.

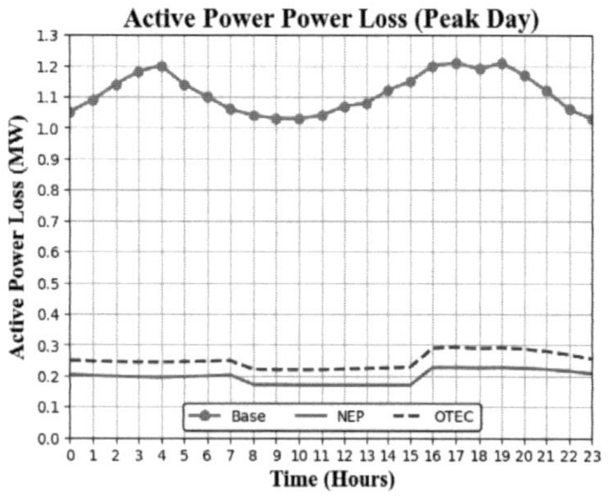

Figure 2: Hourly Active Power Loss for all three systems

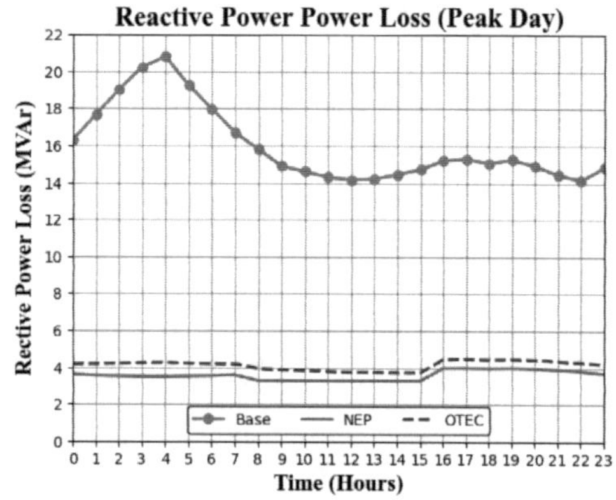

Figure 3: Hourly Reactive Power Loss for all three systems

3.3.1 Power Losses *Figure 2* and *Figure 3* display the active and reactive power loss throughout the day, respectively. As it can be seen, the total active power losses within the system have drastically reduced throughout the day. The base case, NEP solution, and OTEC solution had a total active energy loss of 26.7MWh, 4.8MWh, and 6.1MWh, respectively during the peak day simulation. The decrease in power loss was mainly due to the DG and load being near each other, thus reducing branch losses in the system. Furthermore, the power from nearby DG did not need to travel through multiple transformers, which have higher losses than transmission lines, to service the load. The reactive energy losses saw similar improvements, as the total reactive energy loss decreased from 384.3MVArh (base case) to 86.0MVArh (NEP solution) and 99.3MVArh (OTEC solution). The OTEC solution saw an increase in MVAr compared to the NEP solution as the OTEC solution had less DG and had a more centralized generation.

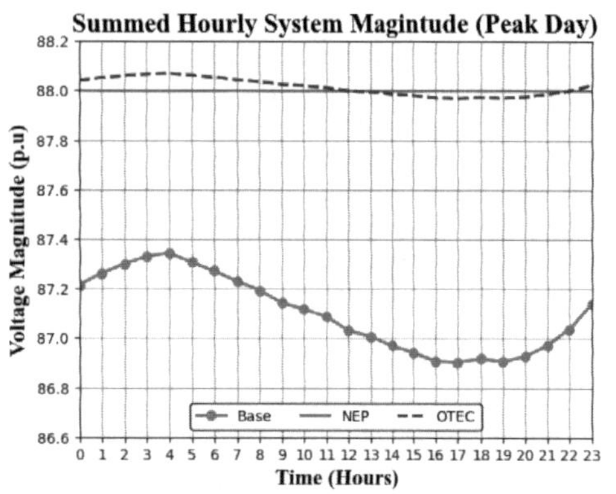

Figure 4: Summed Hourly Voltage Magnitude for all three systems

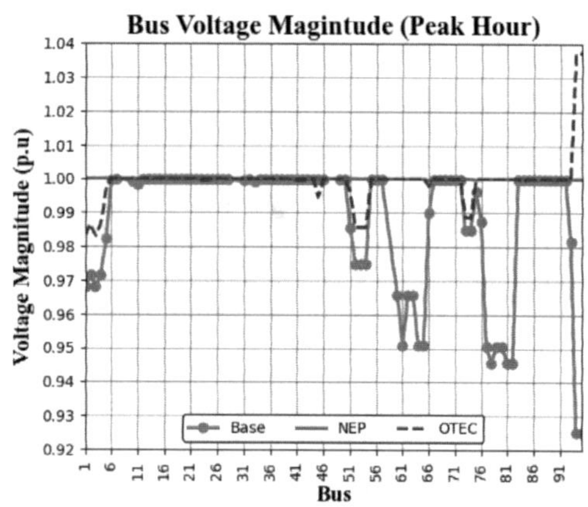

Figure 5: Voltage Magnitude per bus at peak load

3.3.2 Voltage Quality In an ideal scenario, the summed system voltage magnitude would be 88 p.u. *Fig. 4* displays the summed hourly system voltage magnitude during the peak day. The summed voltage for the base case ranged from 86.902 p.u to 87.344 p.u, while for the NEP solution during all hours, the voltage magnitude was approximately 87.999 p.u. However, the OTEC has a slightly wider range of 87.97 p.u to 88.07 p.u. This voltage improvement is due to the amount of voltage support provided by the PV and wind systems located on multiple buses. Additionally, the reduction of PV systems most likely contributed to the OTEC solution's decrease in voltage support. As OTEC was a single source of voltage regulation, the spread of PV installations allowed more buses to have voltage regulation.

The bus voltage magnitude during the peak load (17:00) can be seen in *Fig. 5* for all systems. For the base case, bus 96 experienced the lowest voltage magnitude at 0.9252 p.u. During the peak load, there are 28 buses below 0.99 p.u and 6 below 0.95 p.u. All of the buses below 0.99 p.u are directly connected to or near loads. When implementing the NEP solution, the voltage magnitude during peak load ranged from 0.9992 p.u minimum to 1.0026 p.u maximum. The voltage profile for all buses has vastly improved compared to the base case, and bus 96 now has a voltage of 1.00 p.u.

Unfortunately, for the OTEC solution, the voltage quality of the system during peak load did not drastically improve like the NEP solution. It had some improvements, but bus 96 had a voltage magnitude of 1.037 p.u, an approximate increase of 0.11 p.u. This was due to the OTEC system being installed on bus 96. Buses under 0.99 p.u in the base case have also increased their voltage magnitude in the OTEC solution but remained lower than 1.0 p.u. Both solutions were improvements to the base case, but the voltage for the OTEC solution could be improved further by using reactive power control. Specific solutions are discussed in Section 4.2.

3.3.3 Branch Limits The average daily energy throughout each branch for the base case, NEP and OTEC simulations was 29.73%, 5.16%, and 7.36% of the total MVA branch capacity. There was a significant decrease in the hourly energy flow throughout the branches compared to the base case.

However, examining each system during peak load (17:00) yielded different results. The base case and NEP had no branch violations and a peak branch limit of 97.78% (branch 34) and 57.80% (branch 74), respectively. However, OTEC had a single branch violation at branch 76 (bus 71-72), reaching 106.85% of its limit. This line needs to be upgraded, and specific solutions are discussed in Section 4.2. At the peak load, the average branch limit was 35.61%, 5.24%, and 8.63% for the base case, NEP, and OTEC solution, respectively. Additionally, the base case had 12 lines above 80%, the OTEC solution had 1, and the NEP solution had 0. These improvements were due to the DG being connected closely to the load buses, reducing the energy flow in branches.

3.4 Summary of Results

Table 5 displays a summary of critical metrics between all three systems. Both solutions presented offered better voltage support, lower power losses and decreased the energy flowing through the branches. The NEP solution had the best voltage support and lowest power loss. However, it does have a higher storage amount and higher penetration of DG due to the amount of PV.

Table 5: Key metric comparison between the base case, NEP solution, and OTEC solution

KEY COMPARISONS	BASE CASE	NEP	OTEC
Summed System Voltage Mag. (p.u)	87.10	88.00	88.02
Total Active Energy Loss (MWh)	26.74	4.758	6.02
Total Reactive Energy Loss (MVArh)	384.34	86.06	99.33
DG Peak Power (MW)	0	214	184
Storage Requirements (MWh)	0	803.28	603.17
Peak Load - Bus Voltage Mag. < 0.99	28	0	9
Peak Load - Bus Voltage Mag. > 1.01	0	0	3
Peak Load - Branch MVA Above 80%	12	0	1

4 Discussion

To achieve the outline of respective solutions (OTEC and NEP), there are many technical systems needed. This section discusses the recommended high-level equipment needed for both solutions and their options in providing these systems.

4.1 NEP Technical Requirements
The NEP solution requires 194MW of solar PV, 11MW of wind, 28.5MW of diesel generation, and 9MW of Waste to Energy. Additionally, 194MW of solar PV will follow the IEEE 1547-2018 standard which states that PV-DG must provide some form of voltage regulation [18]. Following the IEEE standard, it was assumed the power factor for the solar installations to be 0.95 instead of 1.0. Although installing 194MW of solar panels onto a small island may seem daunting and an enormous undertaking, using FPV (floating PV) for a large percentage of the installations has many benefits such as higher efficiencies, lower costs, and the possibility of easy dismantling and storage, which is crucial for the hurricane season [19]. Other installation considerations between the PV systems and bus include transformers, DC coupling with storage systems, and other protection systems which can increases losses in the system and can potentially decrease the efficiency of the installed PV systems.

The use of wind turbines in the Caribbean is a complex topic due to the seasonal hurricanes that pass through the area. In 2018, Hurricane Maria passed through Puerto Rico and destroyed an entire wind farm (consisting of 13 turbines rated at 1.8MW each) [20]. Larger standard offshore wind turbines should not be used as they are traditionally not rated for hurricane wind speeds. There is current ongoing research and demonstration projects that are dedicated to typhoon-resistant wind turbines, especially in Asian countries, as they experience the most typhoons and have an extensive wind resource [21]. The typhoon-resistant wind turbines will be a good fit for Grand Cayman if the aforementioned demonstration projects yield adequate results. Any turbine installed should be synchronous to provide voltage support.

The waste-to-energy plant was based on the constraints of a maximum of 9MW and only operating for 8-hour periods. Using day-ahead weather information, the waste-to-energy plant can be scheduled during a specific part of the day to help mitigate the variability of the VRE, similar to the STOR program in the U.K. However, the required human resources must also be flexible. The majority of the current diesel generation can be retired or used as an emergency backup, as three of the largest generation units are more than enough to meet the 28.5MW requirement.

Long-term voltage support may not be required as no bus voltage strayed outside 0.001 p.u of the set point 1.00 p.u throughout the peak day. However, short-term voltage support (time scale of minutes to milliseconds) is required due to the variability of RE systems. Unfortunately, due to scope, a time series analysis could not be done to determine optimum requirements. As the storage solution and PV installations are DC coupled (using the same AC inverter), both systems can offer voltage support when necessary, but the amount of support may be inadequate, and further solutions that can output higher reactive power may be needed.

4.2 OTEC Technical Requirements
All of the aforementioned technical requirements for the NEP solution, such as storage systems, voltage support, frequency support, and specific consideration for generation systems, apply to the OTEC solution. However, the OTEC differs in two key areas, additional voltage support and an OTEC plant. Re-examing *Fig. 5* the entire voltage profile of all buses on the OTEC solution, there was a clear need for voltage support. Both shut capacitors and reactive power controls were used to offer support. Appendix Table 9 identifies key buses outside the specified limits of 0.995 p.u and 1.005 p.u and their respective solutions. To address each type of breach, reactive power needs to decrease to reduce the voltage or vice versa. The reactive power of the OTEC device needs to be controlled to reduce the voltage to acceptable levels. If the voltage magnitude is too low, a shunt switch is required to increase the voltage of that bus. Simple simulations were conducted to approximate the required size of the shunt switch at each bus.

Figure 6 displays the bus voltages during peak load with and without voltage support. For buses 94-96 (above 1.005 p.u), the OTEC reactive power output was decreased. For the other buses (below 0.995), shunt switches were added to increase the voltage. Shunt switches were added manually to increase the voltage. While the solution was not optimal it improved the voltage quality to be within the desired limits. The proposed solution is also not optimum as better solutions use more precise methods and other Flexible AC Transmission Systems (FACTS). However, this does prove that reactive power control for the OTEC device and using shunt switches can improve the voltage quality. Additionally, these solutions are time-dependent and need to be adjusted throughout the day.

As previously identified in Section 3.3.3 when implementing the OTEC solution during the peak load, branch 76 (bus 71-72) had a breach of 106% of its MVA limit. This branch was critical as buses 71 and 72 had 39.9MW of load, approximately shared equally between them. However, the distribution of DG was not equal between buses 71, and 72. Bus 71 had 75MW, and bus 72 had 4MW of installed solar in the

Journal of Physics: Conference Series **2929** (2024) 012011 doi:10.1088/1742-6596/2929/1/012011

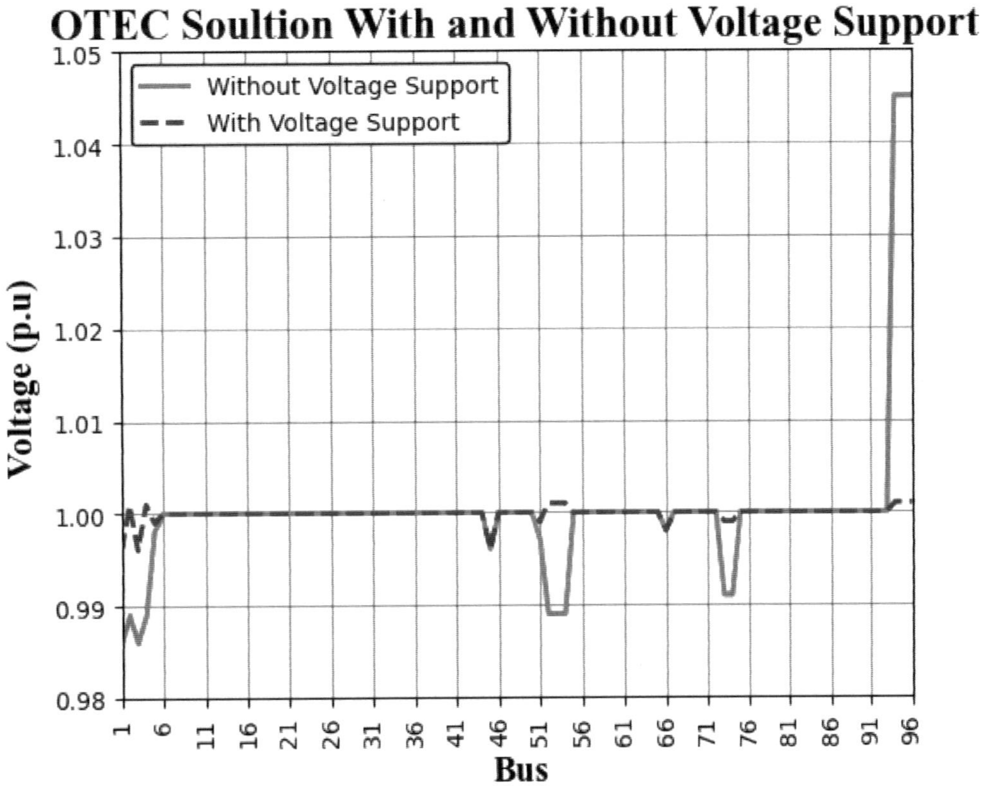

Figure 6: Voltage Magnitude at each bus with and without voltage support for the OTEC solution

OTEC solution. This issue can be solved by either upgrading the connection and secondary emergency connection between the buses or allocating equal VRE between them. Installing an equal amount can reduce the required energy flowing from bus 71 to bus 72 during peak load. Examining the score of buses 71 and 72 during the GSA yields a slight difference of $<1 \times 10^{-3}$, so spreading solar PV installations would have a minimal impact. Further evidence was in the NEP solution, where buses 71 and 72 had a similar solar capacity, but branch 76 (bus 71-72) reached a maximum of 12.54% of its MVA limit. In future research, another power quality index monitoring branch load can be added into the GSA to avoid breaching branch limits.

A 15MW OTEC is required for this solution which is not as simple as installing a 15MW solar plant. These two technologies are entirely different in operation, size, cost, and maturity. Installing a 15MW OTEC plant would be one of the world's first OTEC plants of that size. However, OTEC has a high probability of working and providing a significant portion of the baseload. A more cautious approach would be to install a smaller <1MW plant to determine the feasibility of such an investment.

Recall that the NEP aims to integrate the desired portfolio by 2037. Thus there is room for integrating starter technology and switching to more efficient and better technology later. Demonstration projects must be conducted as soon as possible to avoid technology lock-in. For instance, if redux flow batteries are the optimal storage solution when the installation starts, but other technologies such as liquid metal batteries become a feasible alternative during the installation, switching to liquid metal batteries is recommended to avoid redux flow batteries lock-in. Avoiding technology lock-in is crucial for efficiency and sustainability, especially for storage solutions.

Finally, the Cayman Islands must closely observe the world's technological advances to use the best technology as soon as possible. There are many advances not discussed in this paper that may play a crucial part in the future.

4.3 Storage Requirements

The variety of storage possibilities in Grand Cayman is large, ranging from established technologies such as Lithium-Ion batteries to newer more sustainable options such as Pumped Heat Energy Storage (PHES). However, not all should be used at the same level. Lithium-ion batteries are undoubtedly the most popular form of energy storage globally for large and small-scale applications. However, these storage methods for long-term and large-scale support are not sustainable due to the materials used and costs. Their installation should be kept at a minimum and only used for grid support, and fast energy storage applications in remote areas. More sustainable storage solutions such as PHES and liquid metal batteries are less mature but should be considered and global projects should be closely monitored to determine their feasibility [22], [23]. Unfortunately, due to the flat elevation of the Cayman Islands more mature energy storage solutions such as pumped hydro storage are not feasible as there are no natural reservoirs with large height differences.

The top storage solutions are demand-side management (DSM) and redox flow batteries. DSM is potentially the easiest and most effective way to manage the system energy flow, especially if used on all types of air conditioning and large-scale refrigeration. Using DSM would allow the system operators to reduce the load for hours, effectively acting as virtual energy storage with immediate and almost 100% efficient results [24]. There are control methods, strategies, and optimizations designed to efficiently use smart thermostats to remotely measure and change the temperature set-point of the thermostat [24], [25], [26]. Increasing the temperature remotely of homes and commercial businesses by a few degrees can drastically decrease the load on request. There are standard solutions from companies like NEST which provide central control of participating devices [27]. Although it may be the easiest and cheapest to implement, it involves participation from the public, which may be hesitant.

Flow batteries are a relativity new technology that has been identified by IRENA as a candidate for energy shifting/load levelling, and enhanced frequency response [29]. A recent demonstration completed by Yano, Hayashi, Kumamoto et al. in Hokkaido, Japan, showed their storage systems allowed for fast frequency response and allowed for outdoor installations that are stackable which can increase their energy density per square meter [28]. Although flow batteries may cost more than other battery systems, it has the biggest potential for long-term applications, safety, and future uses [28].

Other new technologies include vehicle to grid storage (V2G) where the utility and private citizens have a pre-established agreement in which the utility can draw the stored energy from a parked electric vehicle's battery. Using these programs would allow the grid to become more stable and reliable. The challenges of implementing such solutions include public participation, bi-directional converters, and a strong I.T infrastructure to measure storage levels, communicate between devices, and compute optimum times for recharge of the storage systems [30], [31].

In summary, for both implemented solutions, the following ranked list of energy systems should be used: demand-side management, redox flow batteries, V2G, liquid metal batteries, pumped heat storage, and lithium-ion batteries.

4.4 Storage Optimization

A further optimization level can be completed by adjusting the output schedule of both waste-to-energy, and conventional generation. Having these types of generation on only during the night reduces the storage requirements while staying within the NEP's goals.

Two improved storage solutions are proposed, with each having its advantages and disadvantages. These solutions are not guaranteed optimal solutions, but they do reduce the storage capacity. The results of these solutions can be seen in *Fig. 7*. Using waste-to-energy and diesel generation only at night reduces the storage requirements from 803.28MWh to 532.23MWh for the NEP solution and 603.17MWh to 344.92MWh for the OTEC solution. Optimizing for storage capacity also reduces the required ramp rate. Consequently, the power capacity for diesel generation has increased from 28.5MW to 54MW. The increase in diesel generation at night will increase the losses and reduces the voltage quality as the storage systems near loads are no longer dominant providers. More units need to also be in service to cover primary units that are in long-term maintenance.

Table 6 displays the total storage required and peak charge rate for all solutions presented. An in-depth analysis of the storage solution was out of scope, but there is a possibility for more complex and extensive storage optimization as this paper did not consider ramp rates, discharge cycles, optimal locations, depth of discharge or distributed storage (such as V2G or virtual grids) when optimizing.

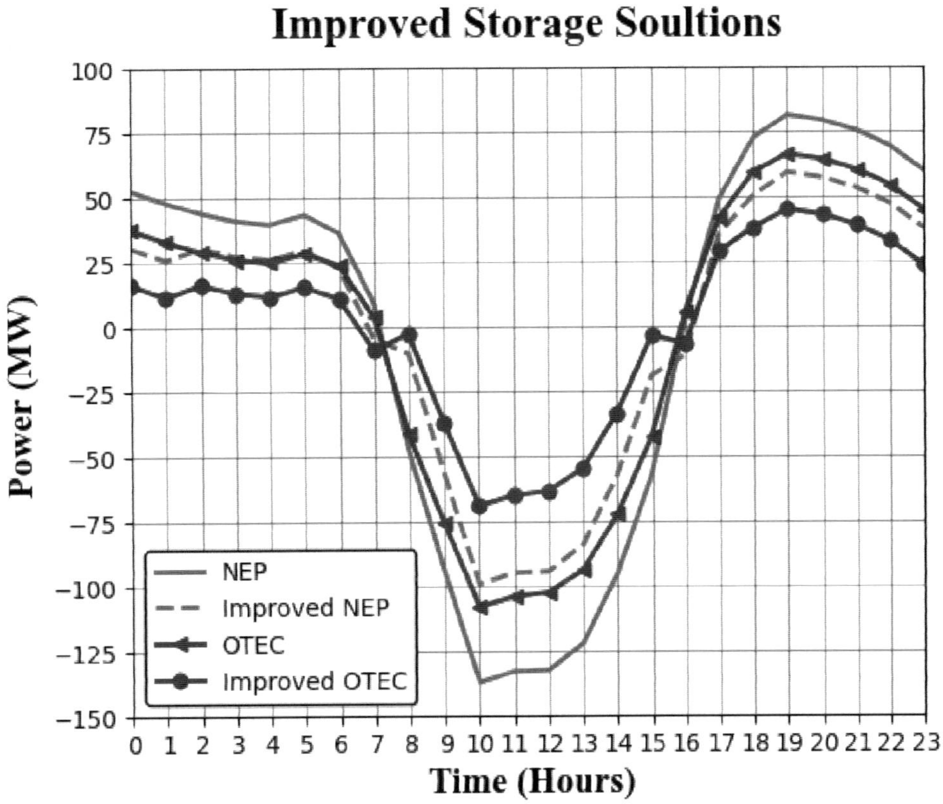

Figure 7: Storage improvements for each solution

Table 6: Comparison of storage solutions

STORAGE SOLUTION	NEP	OTEC	IMP. NEP	IMP. OTEC
Total Storage Required (MWh)	803.28	603.17	532.23	344.92
Peak Discharge Rate (MW)	81.41	66.39	59.30	45.07

4.5 Further Research

Before conducting further research, general improvements to the algorithm and model should include:

- An index that accounts for branch limits
- Newly installed substations and solar plant
- Accurate load information per feeder
- Reactive power support

There are two general directions for further research that operate on different time scales. The first general direction is widening the scope and the simulation time scale by including economic and environmental impacts indices and modelling the entire year to capture seasonal changes. The second general direction is to decrease the time scale from hourly to seconds to examine other power quality challenges. Attention would be given to voltage and frequency support by optimizing the size and location of reactive power compensators/FACTS devices and fast energy storage systems.

5 Conclusion

The Grand Cayman transmission grid was modelled on an hourly basis. Two solutions were proposed using a Greedy Strategy Algorithm. The first solution involved high penetration of PV (NEP solution), and the second included a 15MW OTEC plant used as a baseload (OTEC solution). The algorithm was built so that multiple different types of DG's could be injected into the grid with different power factors, increments and maximum power outputs. The solution's voltage quality, active power losses, reactive power injection, branch limits and storage requirements were compared and discussed. The NEP solution required more storage and solar but had a better voltage profile and fewer losses than the OTEC solution. Although reducing the storage requirements, the OTEC solution required shunt switches and reactive power output control of the OTEC plant to improve its voltage quality. Rudimentary improvements on both storage and voltage quality were conducted. However, in-depth optimization can be completed to improve these systems further. Regardless, both solutions are an improvement compared to the base case and simultaneously show a potential way to reach 70% renewable generation.

6 Acknowledgements

J.A. Ebanks would like to express his deepest gratitude to Alma Lopez, Kody Ebanks, Kyle Ebanks, and their parents.

7 Appendix

Table 7: The GSA's generation results with corresponding DG type, bus location, and size for the OTEC Solution

BUS	MVA BASE	PMAX	QMAX	TYPE	OF TOTAL DG	OF TOTAL LOAD
96	18.75	15	11.25	OTEC	8.15%	7.40%
65	11.25	9	6.75	WtE	4.89%	6.32%
82	14.74	14	4.60	Solar	7.61%	5.43%
79	17.89	17	5.59	Solar	9.24%	6.36%
63	9.47	9	2.96	Solar	4.89%	3.80%
81	11.58	11	3.62	Solar	5.98%	4.39%
62	8.42	8	2.63	Solar	4.35%	3.63%
72	4.21	4	1.31	Solar	2.17%	18.98%
93	2.11	2	0.66	Solar	1.09%	0.00%
80	5.26	5	1.64	Solar	2.72%	2.20%
71	78.95	75	24.65	Solar	40.76%	20.18%
75	1.05	1	0.33	Solar	0.54%	0.00%
25	2.11	2	0.66	Solar	1.09%	0.00%
65	1.05	1	0.33	Solar	0.54%	0.00%
71	1.25	6	0.75	Wind	3.26%	0.00%
82	1.25	1	0.75	Wind	0.54%	0.00%
62	1.25	1	0.75	Wind	0.54%	0.00%
79	1.25	2	0.75	Wind	1.09%	0.00%
81	1.25	1	0.75	Wind	0.54%	0.00%

Table 8: The GSA's generation results with corresponding DG type, bus location, and size NEP Solution

BUS	MVA BASE	PMAX	QMAX	TYPE	OF TOTAL DG	OF TOTAL LOAD
96	11.250	9	6.750	Waste	4.96%	7.40%
82	13.684	13	4.273	Solar	6.04%	5.43%
65	16.842	16	5.259	Solar	7.43%	6.32%
79	16.842	16	5.259	Solar	7.43%	6.36%
63	8.421	8	2.629	Solar	3.71%	3.80%
3	8.421	8	2.629	Solar	3.71%	3.00%
54	7.368	7	2.301	Solar	3.25%	2.77%
4	2.105	2	0.657	Solar	0.93%	1.03%
96	1.053	1	0.329	Solar	0.46%	0.00%
74	1.053	1	0.329	Solar	0.46%	1.13%
95	7.368	7	2.301	Solar	3.25%	5.52%
64	10.526	10	3.287	Solar	4.64%	4.41%
81	10.526	10	3.287	Solar	4.64%	4.39%
62	8.421	8	2.629	Solar	3.71%	3.63%
72	30.526	29	9.532	Solar	13.46%	18.98%
53	6.316	6	1.972	Solar	2.79%	2.58%
80	3.158	3	0.986	Solar	1.39%	2.20%
71	48.421	46	15.119	Solar	21.36%	20.18%
45	1.053	1	0.329	Solar	0.46%	0.88%
93	1.053	1	0.329	Solar	0.46%	0.00%
1	1.053	1	0.329	Solar	0.46%	0.00%
71	1.250	3	0.750	Wind	0.55%	0.00%
65	1.250	1	0.750	Wind	0.55%	0.00%
82	1.250	1	0.750	Wind	0.55%	0.00%
72	1.250	1	0.750	Wind	0.55%	0.00%
79	1.250	1	0.750	Wind	0.55%	0.00%
63	1.250	1	0.750	Wind	0.55%	0.00%
53	1.250	1	0.750	Wind	0.55%	0.00%
64	1.250	1	0.750	Wind	0.55%	0.00%
81	1.250	1	0.750	Wind	0.55%	0.00%

Table 9: Voltage support solutions for the OTEC simulation

BUS	VOLTAGE BEFORE	VOLTAGE AFTER	SUPPORT SOLUTION
1	0.98640	0.99597	0.55 MVAR (Shunt Switch)
2	0.98921	1.00060	0.30 MVAR (Shunt Switch)
3	0.98639	0.99596	0.55 MVAR (Shunt Switch)
4	0.98921	1.00060	0.30 MVAR (Shunt Switch)
52	0.98880	1.00141	0.50 MVAR (Shunt Switch)
53	0.98880	1.00141	0.50 MVAR (Shunt Switch)
54	0.98880	1.00140	0.50 MVAR (Shunt Switch)
73	0.99092	0.99897	0.25 MVAR (Shunt Switch)
74	0.99092	0.99897	0.25 MVAR (Shunt Switch)
94	1.04537	1.00139	Decrease OTEC MVAR Output
95	1.04536	1.00138	Decrease OTEC MVAR Output
96	1.04539	1.00140	Decrease OTEC MVAR Output

References

[1] A. Pathak, P. E. Van Beynen, F. A. Akiwumi, and K. C. Lindeman, "Impacts of climate change on the tourism sector of a Small Island Developing State: A case study for the Bahamas," *Environmental Development*, vol. 37, Mar. 2021, pp. *1–13,* DOI:10.1016/*j.envdev.2020.*100556.

[2] T. Hodge, and E. Fasching, "During 2021, U.S. retail electricity prices rose at fastest rate since 2008." Energy Information Administration. USA, Tech, Rep., Mar. 1, 2022. [Online]. Available: https://www.eia.gov/todayinenergy/detail.php?id=51438

[3] Caribbean Utilities Company, "Caribbean Utilities Company 2021 Annual Report," Caribbean Utilities Company. George Town, Grand Cayman, Fin. Rep., Feb. 11, 2022. [Online]. Available: https://www.cuc-cayman.com/financial-archive

[4] S. Patterson, and S. Goldfarb, "Why are gasoline prices so high? Ukraine-Russia war sparks increases across U.S," Wall Street Journal. New York City, New York, USA, Apr. 1, 2022, [Online]. Available: https://www.wsj.com/articles/why-gas-prices-expensive-11646767172

[5] Caribbean Utilities Company, "Caribbean Utilities Company 2022 First Quarter Report", Caribbean Utilities Company. George Town, Grand Cayman, Fin. Rep., May 4, 2022. [Online]. Available: https://www.cuc-cayman.com/financial-archive

[6] The Government of The Bahamas, "The Bahamas National Energy Policy", The Government of The Bahamas, Nassau, The Bahamas, 2013. [Online]. Available: `https://www.climate-laws.org/geographies/bahamas-the/policies/national-energy-policy-2013-2033`

[7] Cayman Islands Government, "National Energy Policy 2017-2037", Cayman Islands Government, George Town, Grand Cayman, Feb. 21, 2017. [Online]. Available: http://www.dlp.gov.ky/portal/pls/portal/docs/1/12374582.PDF

[8] The Economic and Statistics Office, "The Cayman Islands? 2021 Census of Population and Housing Report," The Cayman Islands Government. George Town, Cayman Islands. July, 2021. [Online]. Available: https://www.eso.ky/2021-population-and-housing-census-report.html

[9] Y. V. P. Kumar, and R. Bhimasingu, "Improving resiliency in renewable energy based green microgrids using virtual synchronous machines controlled inverter," in *2015 IEEE Innovative Smart Grid Technologies - Asia (ISGT ASIA)*, 2015, pp. *1–6,* DOI:10.1109/*ISGT-Asia.2015.*7387178.

[10] X. Liang, "Emerging Power Quality Challenges Due to Integration of Renewable Energy Sources," *IEEE Transactions on Industry Applications* vol. 53, no. 2, 2017, pp. *855–866,* DOI:10.1109/*TIA.2016*.2626253.

[11] R. Torabi, A. Gomes, and F. Morgado-Dias, "The duck curve characteristic and storage requirements for greening the island of Porto Santo," in *2018 Energy and Sustainability for Small Developing Economies (ES2DE)*, 2018, pp. *1–7,* 2018 DOI:10.1109/*ES2DE.2018*.8494235.

[12] R. D. Zimmerman, C. E. Murillo-Sanchez, R. J. Thomas, "Matpower: Steady-state operations, planning, and analysis tools for power systems research and education", in *IEEE Transactions on Power Systems*, vol. 26, no. 1, pp. *12–19,* 2011 DOI:10.1109/*TPWRS.2010*.2051168.

[13] R. O. Bawazir, and N. S. Cetin, "Comprehensive overview of optimizing PV-DG allocation in power system and solar energy resource potential assessments", *Energy Reports*, vol. 6, pp. *173–208,* 2020, DOI:10.1016/*j.egyr.2019*.12.010.

[14] E. Ali, S. Abd Elazim, and A. Abdelaziz, "Ant Lion Optimization Algorithm for optimal location and sizing of renewable distributed generations," *Renewable Energy*, vol. 101, pp. *1311-1324,*, 2016, DOI:10.1016/*j.renene.2016*.09.023.

[15] M. Ebanks, "Full Steam Ahead for New Waste System," Mar. 26, 2021. [Online]. Available: `https://www.gov.ky/news/press-release-details/full-steam-ahead-for-new-waste-system`

[16] R. Fujita, A. C. Markham, J. E. Diaz Diaz, J. Rosa Martinez Garcia, C. Scarborough, P. Greenfield, P. Black, and S. E. Aguilera, "Revisiting ocean thermal energy conversion," *Marine Policy,* vol. 36, no. 2, pp. *463–465,*, 2012, DOI:10.1016/*j.marpol.2011*.05.008.

[17] R. Kempener, and F. Neumann, "Ocean Thermal Energy Conversion Technology Brief", Tech. rep., International Renewable Energy Agency (2014). [Online]. Available: `https://www.irena.org/-/media/Files/IRENA/Agency/Publication/2014/`

[18] *IEEE Standard for Interconnection and Interoperability of Distributed Energy Resources with Associated Electric Power Systems Interfaces*, 1547-2018, Feb. 2, 2018.

[19] L. C. Dobre, A. Turcanu, and A. Craciunescu, "Floating Photovoltaic Power Plants," in *International Symposium on Advanced Topics in Electrical Engineering (ATEE)*, vol. 12, pp. *1–4,* 2018, DOI:10.1109/*ATEE52255.2021*.9425257.

[20] A. Kwasinski, "Effects of Hurricane Maria on Renewable Energy Systems in Puerto Rico," in *International Conference on Renewable Energy Research and Applications (ICRERA)*, pp. *383–390,* 2018, DOI:10.1109/*ICRERA.2018*.8566922.

[21] J. Li, Z. Li, Y. Jiang, and Y. Tang, "Typhoon Resistance Analysis of Offshore Wind Turbines: A Review,? *Atmosphere*, vol. 13, no. 3, pp. *451–460*, Mar. 2022, DOI: 10.3390/*atmos*13030451.

[22] A. Benato, and A. Stoppato, "Pumped Thermal Electricity Storage: A technology overview," *Thermal Science and Engineering Progress*, vol. 6, pp. *301–315,* June, 2018, DOI:10.1016/*j.tsep.2018*.01.017.

[23] X. Zhou, H. Zhou, S. Yan, Y. He, W. Zhang, H. Li, K. Wang, and K. Jiang, "Increasing the actual energy density of Sb-based liquid metal battery," *Journal of Power Sources,* vol. 534, 2022, DOI:10.1016/*j.jpowsour.2022*.231428

[24] J. Hao, Y. Zhang, T. T. Ha, and H. Pham T., "Demand Response Scheduling Strategy for Air-conditioning Considering the Users' Somatosensory Compensation," in *2018 International Conference on Power System Technology (POWERCON)*, 2018, pp. 1362–1366. DOI:10.1109/*POWERCON*.2018.8601745, [Online].

[25] Z. Yuqiang, W. Qi, L. Xiaofeng, L. Xueming, Z. Weiguo, and J. Cheng, A Temperature Control Strategy of Fixed Frequency Air Conditioning Loads for Short-Time Peak Clipping, *2020 IEEE 3rd Student Conference on Electrical Machines and Systems (SCEMS)*, pp. *409–414,* 2020, DOI:10.1109/*SCEMS48876*.2020.9352434, [Online].

[26] R. Fan, Y. Li, Y. Cao, W. Xie, Y. Tan, and Y. Cai, An optimization management strategy for energy efficiency of air conditioning loads in smart building, *2016 IEEE 16th International Conference on Environment and Electrical Engineering (EEEIC)*, pp. *1–5*, 2016. DOI:10.1109/*EEEIC.2016*.7555507.

[27] NEST, NEST - Demand Response, 2021. [Online]. Available: https://nest.com/energy-partners/demand-response

[28] K. Yano, S. Hayashi, T. Kumamoto, T. Shibata, K. Yamanishi, and K. Fujikawa, (2017). Development and Demonstration of Redox Flow Battery System. Sumitomo Electric. Hokkaido, Japan. [Online]. Available: https://global-sei.com/technology/tr/bn84/pdf/84-04.pdf

[29] P. Ralon, M. Taylor, A. Ilas, H. Diaz-Bone, and K.-P. Kairies, "Electricity storage and renewables: Costs and markets to 2030,", IREA, Tech. Rep. ISBN 978-92-9260-038-9, Oct. 2017

[30] A. Zangeneh, M. Moeini-Aghtaie, Ali Akbar Moti Birjandi, "EV's vehicle-to-grid implementation through virtual power plants", in *Scheduling and Operation of Virtual Power Plants*, 1st ed. Philadelphia: Elsevier, 2022, ch. 13, sec. 13.4.6, pp. 299–324.

[31] L.Wagner,"Overview of Energy Storage Technologies", in *Future Energy*, 2nd ed. Boston: Elsevier, 2014, ch. 27, sec. 27.5.2, pp. 613–631

James Ebanks received their MEng in Electrical and Electronic Engineering degree from the University of Bristol, United Kingdom, in 2019, and their MSc in Sustainable Energy Systems from The University of Edinburgh, United Kingdom, in 2021.

Daniel Friedrich received their Postgraduate Diploma in Science in Mathematics at the University of Auckland, New Zealand, in 2002, Diplom Technomathematik from the Universitaet Karlsruhe, Germany, in 2005, and their PhD in Applied Mathematics from the University of Southampton, United Kingdom, in 2009. Their current research interest include mathematical modelling of engineering systems, and Thermal Energy Storage.

Simulation of a decentralized floating offshore wind hydrogen production system with hydro-pneumatic energy storage and subsea isobaric hydrogen storage

***Zecheng Zhao*[1]*, Wei Xiong*[1]*, Hu Wang*[1]*, Tonio Sant*[2]*, Rupp Carriveau*[3]*, David S-K Ting*[3] *and Zhiwen Wang*[1*]**

[1] *Department of Mechanical Engineering, Dalian Maritime University, 116026, Dalian, China*
[2] *Department of Mechanical Engineering, University of Malta, Msida MSD2080, Malta*
[3] *Turbulence and Energy Laboratory, University of Windsor, ON N9B 3P4, Windsor, Canada*

*E-mail: wzw@dlmu.edu.cn

Abstract. Hydrogen production using offshore wind power is a promising solution for producing clean fuels in remote areas. However, the intermittency of offshore wind power poses significant challenges to Proton exchange membrane (PEM) water electrolysis systems. Besides, there is generally an imbalance between hydrogen production and hydrogen demand. In this study, hydro-pneumatic electricity energy storage and subsea isobaric hydrogen storage are integrated into the decentralized offshore wind hydrogen production system. The hydro-pneumatic energy storage unit (HPES) is used for mitigating the intermittency and fluctuation of wind power, thereby prolonging the lifespan of PEM electrolyzers. Subsea isobaric hydrogen storage unit is used for replacing conventional isochoric hydrogen storage. In this study, simulation models of a decentralized offshore wind hydrogen production system with various configurations are established with the software Simcenter Amesim 2021.1. The results show that an 83% reduction in the on/off operation can be achieved with the help of hydro-pneumatic electricity energy storage. The isobaric and isothermal storage of compressed hydrogen can be achieved with subsea isobaric hydrogen storage, saving compression energy and facilitating thermal management. In terms of the investigated decentralized offshore wind hydrogen production system, the amount of produced hydrogen is increased by less than 1% by integrating hydro-pneumatic energy storage and subsea isobaric hydrogen storage.

1 Introduction

Climatic anomalies, such as temperature extremes, aridity, and floods, have become increasingly frequent in recent years. Carbon-neutral and net zero emissions must be given more attention for future generations [1-3]. Currently, renewable energy is the most accessible and promising solution, particularly solar energy and wind [4]. In recent years, floating offshore wind in deep water has drawn great attention due to the enormous potential and fewer interest conflicts. However, the energy loss and initial cost for transmitting electricity generated from floating offshore wind farms in remote and deep seas are significantly high [5,6]. Besides, offshore wind remains fluctuating and intermittent. In recent years, green hydrogen production has been identified as a promising solution for developing offshore wind in remote and deep seas [7,8].

Hydrogen is an energy carrier with a high mass energy density (142 MJ/kg) and is a significant industrial chemical [9]. It is also the core of Power-to-X. For accommodating the

Journal of Physics: Conference Series **2929** (2024) 012012 doi:10.1088/1742-6596/2929/1/012012

ntermittency and volatility of wind power, the Proton exchange membrane (PEM) electrolyzer is widely accepted for offshore wind hydrogen production owing to its relatively mature technical state and quick response dynamics. However, highly dynamic and frequent on/off operation can markedly degrade the electrolyser lifetime [10-12]. Intermittent on/off operations can lead to a decrease in the activity of the catalyst layer and an enhancement of Iridium dissolution at the anode active sites. The degradation is accelerated by the complete shutdown of the PEM electrolyzer at short intervals during dynamic operation [13]. Beyond this, the BOP (balance of plant) cannot be reliably operated with fluctuating and intermittent wind power. Intermediate energy storage located between the generation source and the electrolyser is a promising solution for solving these problems [14]. The load profiles of the PEM electrolyzer can be smoothed and regulated with energy storage, thereby mitigating the fluctuation and reducing the on/off operation. Currently, Li-ion battery energy storage is the commercially available option that could be deployed on the offshore platform. However, the safety issue is an important challenge. The harsh marine environment also challenges the lifetime and economy of Li-ion battery energy storage systems [10]. In addition, in recent years, some other new energy storage methods are also developing rapidly, such as underwater CAES and buoyancy energy storage [15-18]. Similarly, hydrogen storage is essential for bridging the imbalance between hydrogen production and hydrogen consumption. Large steel pressure vessels are generally used for storing high-pressure compressed hydrogen. Safety is yet another important challenge, especially when the hydrogen storage systems have to sustain the complex harsh marine loads on unmanned offshore platforms. These defects can be avoided if the energy storage systems, including electricity energy storage and hydrogen energy storage, can be deployed on the seabed.

Underwater energy storage coupling hydro-pneumatic energy storage with subsea isobaric hydrogen storage may provide a viable solution to addressed the above-mentioned problems of floating offshore wind hydrogen production. Underwater energy storage has received extensive attention in recent years because of various advantages [18-22]. Settino et al. [10] proposed a hydro-pneumatic energy storage system that can be used as an interconnection between the fluctuating wind electricity and the electrolyser. The simulation results showed that the hydro-pneumatic energy storage system provides a stabilized and green power supply. Short-term fluctuations of the power input to the electrolyzer could be avoided and a reduction of the On/Off cycles could be obtained. Borg et al. [23] involved the numerical simulation of an offshore Hydro-Pneumatic Energy Storage system. The effect of this system on smoothing power fluctuations of offshore wind power was analysed. Astolfi et al. [24] conducted a case study where the UW-CAES is coupled to a wind farm for peak shaving and dispatchability enhancement. The impacts of a realistic power input on performance and plant flexibility were evaluated. The above research mainly focuses on the applications of underwater energy storage systems to alleviate the intermittency and volatility of offshore wind power. This has an important influence on the lifespan of the PEM electrolyzer. Underwater hydrogen also plays an important role in alleviating the imbalance between hydrogen production and consumption. Dinh et al. [25] integrated subsea hydrogen geological storage with offshore wind farms. Hassanpouryouzband et al. [26] analyzed the prospects and scientific challenges in subsea hydrogen geological storage. In addition, the economic analysis is also very important. Wang et al. [27] proposed a OWTs-UWCHES (Offshore Wind Turbines & Underwater Compressed Hydrogen Energy Storage) concept. The simple payback period (SPB), discounted payback period (DPB), net present value (NPV), and levelized cost of hydrogen (LCOH) are investigated to evaluate the economic performance of the system. The one-at-a-time (OAT) local sensitivity analysis is conducted to determine the influence of various parameters on the economic performance. Overall, the proposed OWTs-UWCHES concept is viable. Other economic analyses of hydrogen production at offshore are also included [28-30]. Overall, underwater hydrogen storage technologies are still in the conceptual stage and literature related to research on subsea isobaric hydrogen storage is still lacking.

This study presents a decentralized offshore wind hydrogen production system integrating both hydro-pneumatic energy storage and a subsea isobaric hydrogen storage, to concurrently reduce the number of electrolyzer On/Off cycles and address the supply-demand mismatch related to hydrogen production. A numerical model to simulate the operational characteristics is implemented using the software Simcenter Amesim 2021.1. This paper is organized as follows: In Section 2, two different configurations of modelled floating offshore wind hydrogen production systems are introduced. The simulation models are established in Section 3. In Section 4, the performances of two systems are compared and analyzed. Finally, the conclusion is made in Section 5.

2. System Description

The basic configuration of a conventional hydrogen production system powered by offshore wind is shown in Figure 1 (a). This configuration is called 'the conventional system' in this study. The system mainly consists of three basic subsystems: a floating offshore wind turbine, a PEM water electrolysis and a high-pressure isochoric hydrogen storage. The floating wind turbine supplies electricity for PEM water electrolysis system. The produced hydrogen is compressed to a high pressure and stored in metal pressure vessels. The stored high-pressure hydrogen is then offloaded to a vessel. Figure 1 (b) shows the basic configuration of the proposed decentralized floating offshore wind hydrogen production system integrating a hydro-pneumatic energy storage and a subsea isobaric hydrogen storage. In this study, this configuration is being referred to as 'the proposed system'. In this system, excess electricity generated by offshore wind turbine is stored by the hydro-pneumatic electricity energy storage unit. It can be used to compensate for power shortages when offshore wind power is insufficient for hydrogen production. The subsea isobaric hydrogen storage unit replaces conventional isochoric metal high-pressure hydrogen storage tanks. The subsea isobaric hydrogen storage unit uses the hydrostatic pressure of seawater to keep the pressure of stored hydrogen constant, and the storage pressure increases with the increase of depth. A composite membrane is added to the unit to prevent hydrogen from dissolving in seawater. In the energy storage process, compressed hydrogen inlets the flexible bladder through the hydrogen transmission pipeline, causing the volume of stored hydrogen to increase. The seawater discharges from the storage tank through outlet ports. In the energy release process, the volume of compressed hydrogen in the flexible bladder gradually decreases. The seawater continuously flows into the storage tank through the inlet. In addition, an energy management system of subsea hydro-pneumatic energy storage is important to activate and control the On/Off switching of the energy storage process. The variable speed of the electric motor on the energy storage side and the variable displacement of the hydraulic motor on the energy release side are realized by PID control. In the following section, simulation models of both systems are established.

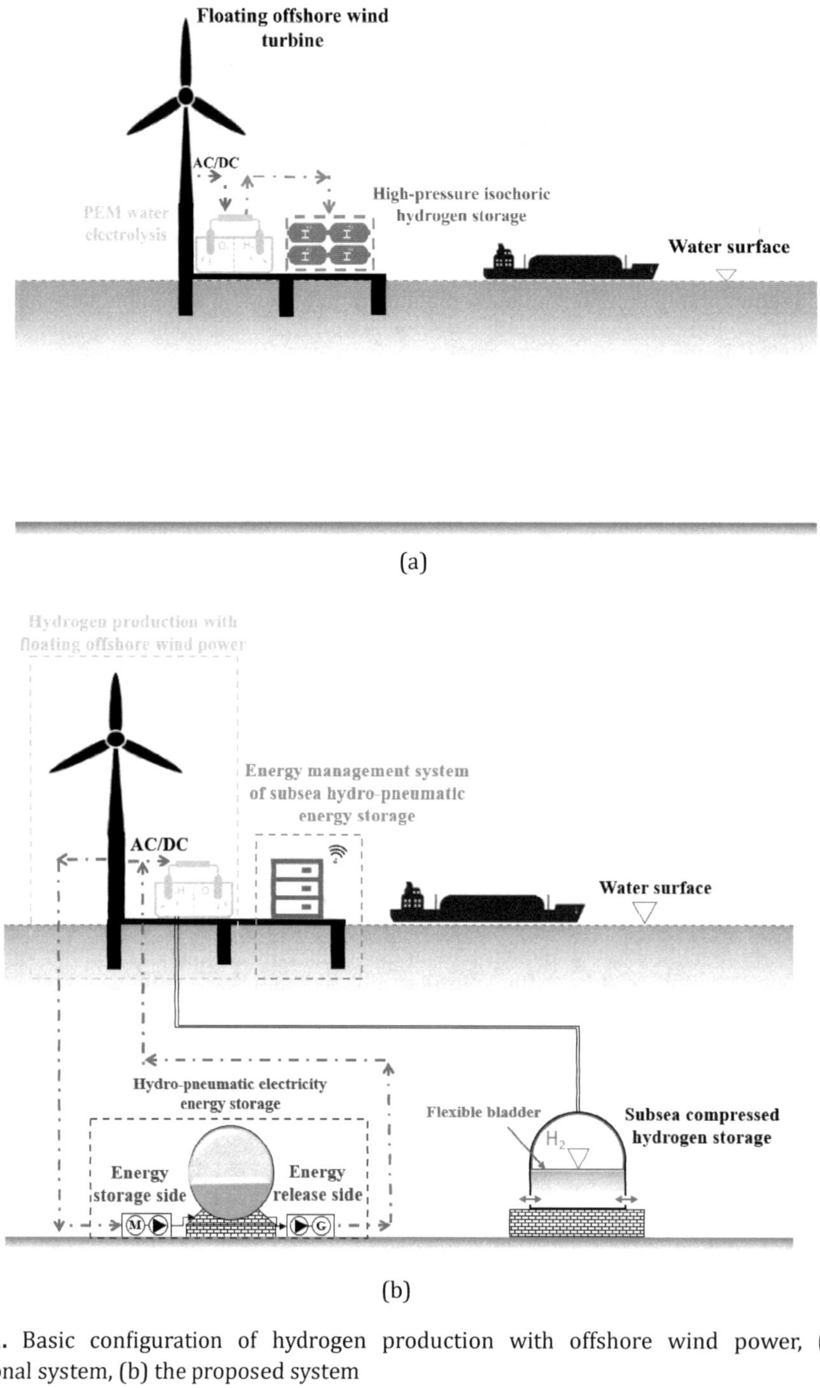

Figure 1. Basic configuration of hydrogen production with offshore wind power, (a) the conventional system, (b) the proposed system

3. System mathematical model and simulation model

The simulation models of the two systems shown in Fig.2 are established by integrating the various components in different libraries of the software Simcenter Amesim 2021.1[31]. The following assumptions are made for simplifying the system model.

- The wind turbine power coefficient is constant.
- It does not take into account the energy loss in the transmission of electricity from offshore wind turbine to PEM electrolyzer.
- The variation of physical properties such as density and viscosity of seawater with temperature is ignored.
- The efficiences of both the pump and motor are assumed to remain constant.
- The energy loss in the process of hydrogen production by PEM electrolyzer is not considered.
- The influence of temperature and pressure on the PEM electrolyzer is ignored.
- Hydrogen is considered to be an perfect gas.

3.1 Mathematical model

(1) Offshore wind power

The main parameter of a 15MW wind turbine is shown in Table 1 [32].

Table 1. The main parameter of 15MW wind turbine.

Parameters	Unit	Value
Rotor diameter D	m	240
Rated power P_m	MW	15
Cut-in wind speed v_{in}	$m \cdot s^{-1}$	3
Rated wind speed v_n	$m \cdot s^{-1}$	10.56
Cut-out wind speed v_{out}	$m \cdot s^{-1}$	25
Wind turbine power coefficient C_p	/	0.46

The output power of the wind turbine is calculated as follows [33].

$$P = \begin{cases} 0, & 0 < v < 3 \\ \frac{1}{2} C_p \rho \pi R^2 v^3, & 3 \le v < 10.56 \\ P_m, & 10.56 \le v < 25 \\ 0, & v \ge 25 \end{cases} \tag{1}$$

where C_p is wind turbine power coefficient, ρ is air density, R is rotor radius, v is wind speed, P_m is rated power. Figure 2 shows the power characteristic curve of 15 MW wind turbine.

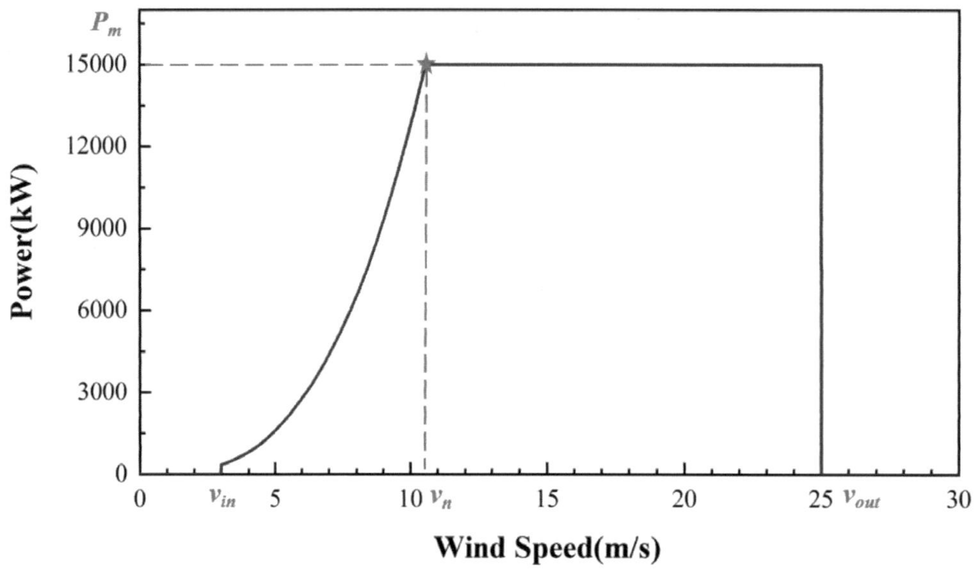

Figure 2. The power curve of 15 MW wind turbine (v_{in}-cut-in wind speed, v_n-rated wind speed, v_{out}-cut out wind speed)

(2) PEM electrolyzer

The PEM electrolyzer voltage is the sum of thermo-neutral voltage, activation overpotentials, ohmic overpotentials and mass transfer overpotentials. The mass transfer losses are not significant for low and moderate current densities if the flow field is appropriate for gas removal. Thus, the mass transfer overpotentials can be neglected for up to 3 A/cm². These are not considered in this study [34]. The PEM electrolyzer output voltage is shown as follows,

$$\begin{cases} U_{el} = V_{el,oc} + V_{el,act} + V_{el,ohm} \\ V_{el,oc} = \frac{\Delta G_f}{2F} + \frac{RT_{el}}{2F}[\ln(\frac{p_{H_2}p_{O_2}^{0.5}}{a_{H_2O}})] \\ V_{el,act} = \frac{RT_{el}}{\alpha_{an}F}\text{arcsinh}\left(\frac{i}{2i_{0,an}}\right) + \frac{RT_{el}}{\alpha_{ca}F}\text{arcsinh}\left(\frac{i}{2i_{0,ca}}\right) \\ V_{el,ohm} = \frac{\delta_m}{\sigma_m}i \end{cases} \tag{2}$$

where, σ_m is the membrane conductivity,

$$\sigma_m = (0.00514\lambda_m - 0.00326)\exp[1268(\frac{1}{303} - \frac{1}{T_{el}})] \tag{3}$$

where U_{el} is the output voltage of the PEM electrolyzer, $V_{el,oc}$ is the thermo-neutral voltage, $V_{el,act}$ is activation overpotentials, $V_{el,ohm}$ is the ohmic overpotentials, F is the Faraday constant, ΔG_f is the Gibbs free energy during the reaction process, R is the gas constant, T_{el} is the operating temperature of the PEM electrolyzer, α_{an} is the charge transfer coefficient of anode, α_{ca} is the charge transfer coefficient of cathode, i is the current density, $i_{0,an}$ is the exchange current density of anode, $i_{0,ca}$ is the exchange current density of cathode, δ_m is the membrane thickness, λ_m is the membrane water content, p_{H_2}, p_{O_2} and a_{H_2O} are the effective partial pressures of hydrogen, oxygen and the water activity at the anode in relation to the water electrolysis, respectively.

When multiple electrolyzers are usually used in series, the total output voltage is:

$$U_{elc} = n_c U_{el} \tag{4}$$

where U_{elc} is the total output voltage of PEM electrolyzers, n_c is number of electrolyzers in series.

According to the principles of electrochemistry, the rate of hydrogen production in water electrolysis is directly proportional to the current in the electrolytic circuit. The hydrogen production rate in the PEM electrolyzer can be expressed as [35],

$$q_H = \eta_F n_c \frac{I_{el}}{2F} \tag{5}$$

$$\eta_F = \frac{(I_{el}/S_{el})^2}{K_{f1} + (I_{el}/S_{el})^2} K_{f2} \tag{6}$$

where q_H is the hydrogen production rate, η_F is the current efficiency of the PEM electrolyzer, F is the Faraday constant, K_{f1} and K_{f2} are empirical parameters. The main parameter of PEM electroluzer is shown in Table 2[36].

Table 2. The main parameter of PEM electrolyzer.

Parameters	Unit	Value
Operating temperature T_{el}	K	333.15
Faraday constant F	C/mol	96485
Gas constant R	J/mol ·K	8.314
Gibbs free energy ΔG_f	kJ/mol	23700
Effective partial pressures of hydrogen p_{H_2}	atm	3
Effective partial pressures of oxygen p_{O_2}	atm	3
Water activity a_{H_2O}	/	1
Exchange current density of anode $i_{0,an}$	A/m^2	2×10^{-3}
Exchange current density of cathode $i_{0,ca}$	A/m^2	20
Charge transfer coefficient of anode α_{an}	/	2
Charge transfer coefficient of cathode α_{ca}	/	0.5
Membrane thickness δ_m	m	2.54×10^{-6}
Membrane water content λ_m	/	12.5019
Number of electrolyzers in series n_c	/	1000
Electrolyzer electrode surface area S_{el}	m	0.25
Empirical parameter K_{f1}	/	2.5×10^4
Empirical parameter K_{f2}	/	0.96

(3) Energy management system of subsea hydro-pneumatic energy storage

Before establishing the simulation model of the underwater energy management system, it is necessary to define the error to control the variable speed of the motor on the energy storage side and the variable displacement of the hydraulic motor on the energy release side. The definition of error is,

$$P_{error} = P_e - P_c \tag{7}$$

where P_{error} is power error, P_e is the expected value of power, P_c is the current value of power.

Figure 3 shows the logical control flow diagram of the energy management system. When the output power of wind power P_1 exceeds the value of k_1, the energy storage unit begins to store energy. In order to control the volume and pressure of the accumulator, the maximum pressure value p is set in the system. When the pressure p_1 in the accumulator reaches the maximum pressure, the energy storage unit is not storing energy. On the contrary, the output power of wind power P_1 is lower than the value of k_2, energy storage unit begins to release energy.

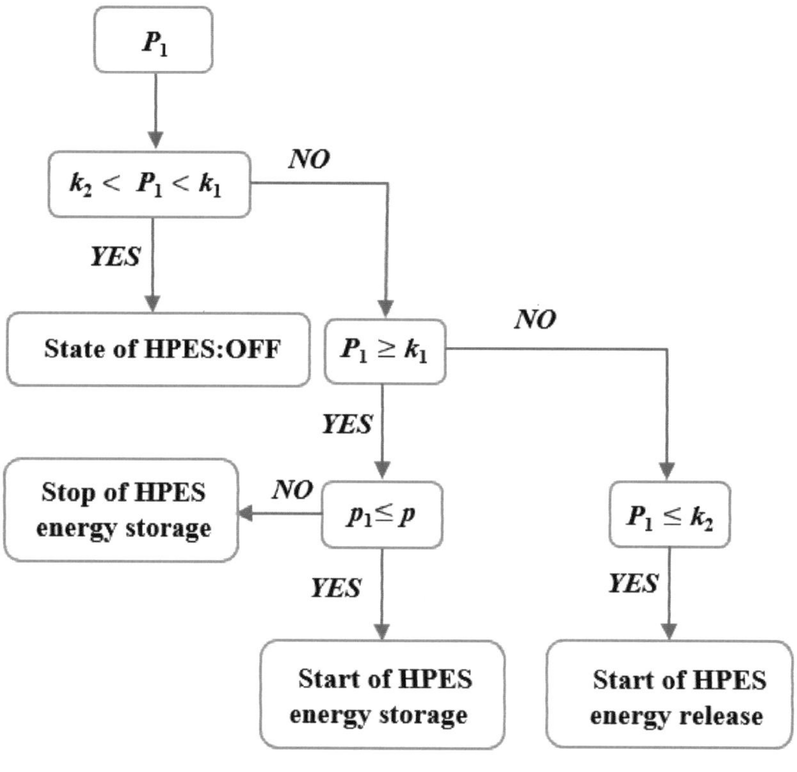

Figure 3. Control logic diagram of the energy management of HPES.

(4) Performance evaluation

When evaluating the performance of the system, the system efficiency and energy density are often used as indicators. In this study, the power, accumulated energy and efficiency are defined as shown in Table 3. In order to better understand the different power terms across the both the conventional and proposed systems, a schematic diagram is drawn, as shown in Figure 4.

Table 3. The definition of the power, accumulated energy and efficiency in the system.

Power	Symbol	Accumulated energy	Efficiency/utilization rate	Symbol
Output electrical power of wind turbine (Input electrical power of the electrolyzer in the conventional system)	P_1		Electrical energy efficiency of hydrogen production with offshore wind power system (Electrical energy utilization rate of the electrolyzer in the conventional system)	$\eta_1 = \frac{W_3}{W_1}$
Input electrical power of the electrolyzer in the proposed system Electrical power consumed by the electrolyzer	P_2		Electrical energy utilization rate of the electrolyzer in the proposed system	$\eta_2 = \frac{W_3}{W_2}$
Electrical power consumed by the electrolyzer	P_3	$W_i = \int_{t_0}^{t_1} P_i\, dt$		
Energy storage power of hydro-pneumatic energy storage unit	P_4		Efficiency of the energy storage side	$\eta_3 = \frac{W_4}{W_5}$
Input electrical power of the energy storage side	P_5		Efficiency of the energy release side	$\eta_4 = \frac{W_7}{W_6}$
Energy release power of hydro-pneumatic energy storage unit	P_6		Efficiency of hydro-pneumatic energy storage unit	$\eta_5 = \frac{W_7}{W_5}$
Output electrical power of the energy release side	P_7			

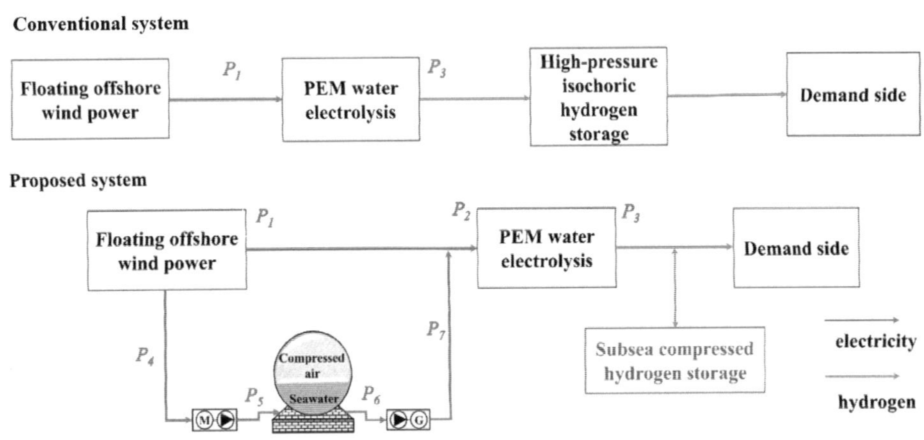

Figure 4. Two different systems of power flow diagrams.

For the hydro-pneumatic energy storage unit, the electrical energy density is calculated as follows

$$ED_{elec} = \frac{W_7}{V_0} \tag{8}$$

where ED_{elec} is electrical energy density and V_0 is initial air volume of energy storage device.

3.2 Simulation model

The simulation models for both the conventional and proposed decentralised offshore wind-driven hydrogen production systems (Figure 1) were implemented using software Simcenter Amesim 2021.1. The schemetic diagrams for the two models are presented in Figure 5. The hydro-pneumatic energy storage system consists of an electric motor, pump, hydraulic motor, generator, and a hydro-pneumatic energy accumulator. It is divided into two energy processes: charging and discharging. During system charging, offshore wind power drives the electric motor to operate the seawater pump, pressurizing seawater into the energy storage device to compress the gas for energy storage. During system discharging, the compressed air expands, forcing the seawater to exit the accumulator and drive the hydraulic motor to generate electricity through the generator, supplying power to the electrolyzer. It should be mentioned that, the maximum pressure is 35 MPa and multiple hydrogen storage tanks can store hydrogen in turn through logic control in conventional isochoric hydrogen storage. The temperature is 293.15 K in the atmospheric environment. For hydro-pneumatic energy storage and subsea isobaric hydrogen storage, the water depth of storage is 500 m, and the environmental temperature and pressure is 283.15 K and 5 MPa, respectively. Therefore, the constant pressure of subsea isobaric hydrogen storage is 5 MPa. For the isobaric hydrogen storage system, it uses the hydrostatic pressure of seawater to achieve a constant hydrogen storage pressure. In the simulation model, subsea isobaric hydrogen storage is equivalent to a vertically arranged pneumatic cylinder model. The mass load is used for simulating the gravity force of water column, thereby achieving the constant-pressure characteristic. Then the heat transfer coefficient is considered in two different hydrogen storage models. The heat transfer coefficient between hydrogen in the conventional isochoric hydrogen storage and air was assumed to be 5 $W/m^2{\cdot}K$, while the heat transfer coefficient between hydrogen in the subsea isobaric hydrogen storage and seawater was taken equal to 10 $W/m^2{\cdot}K$. In addition, the wind speed data comes from the measured value. The main component parameters in simulation models are presented in Table 4.

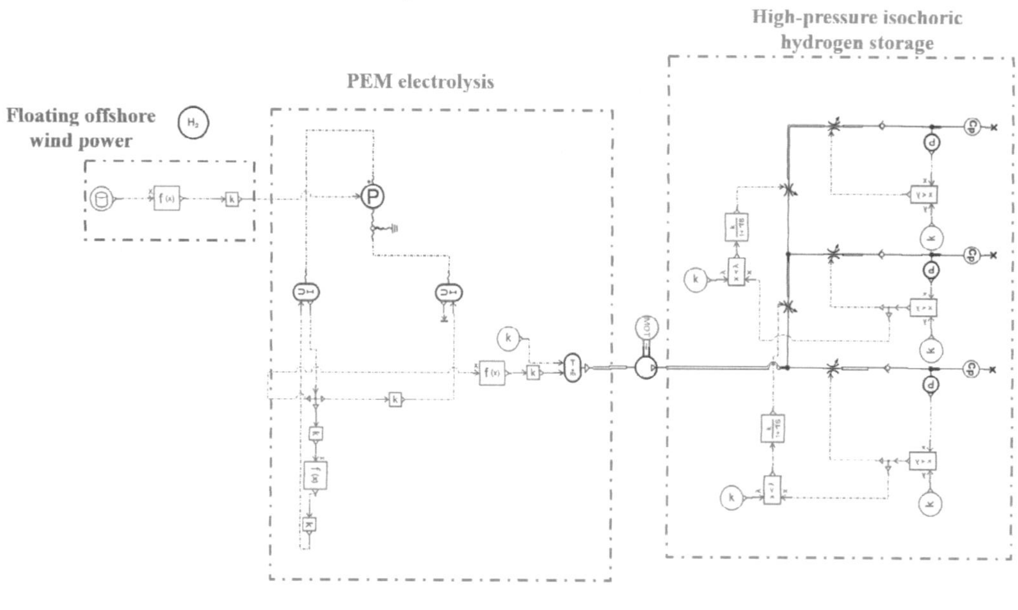

(a)

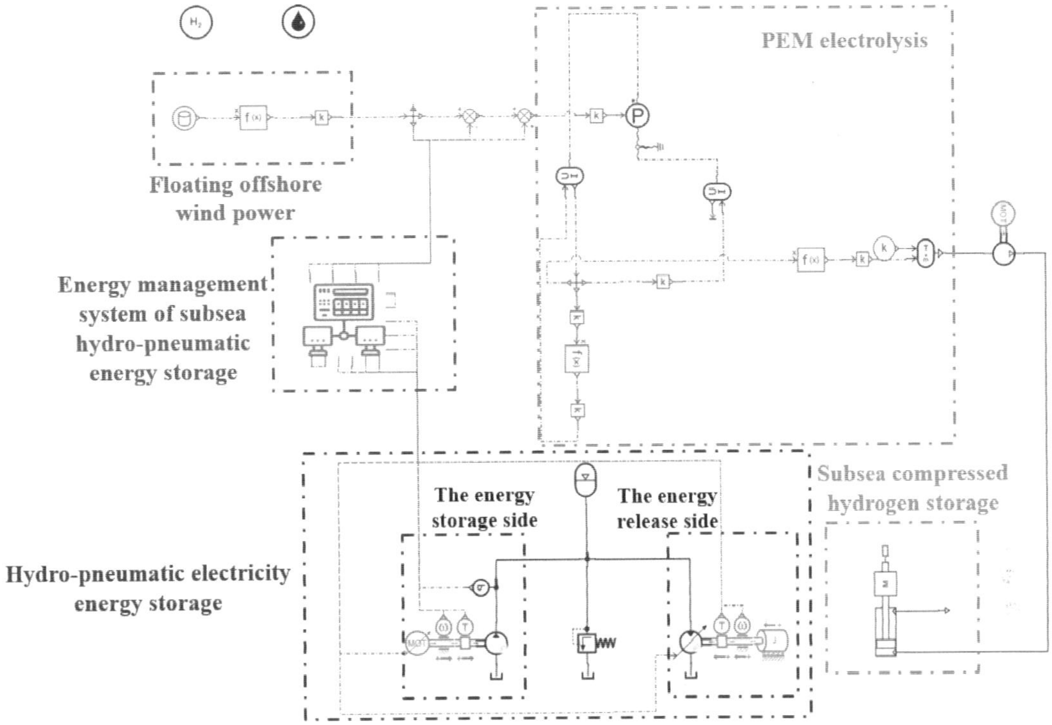

(b)

Figure 5. The simulation model, (a) conventional hydrogen production with offshore wind power system, (b) the decentralized offshore wind hydrogen production system with hydro-pneumatic energy storage and subsea isobaric hydrogen storage.

Table 4. The main component parameters in simulation model.

Parameters	Unit	Column heading
Maximum working current of electrolyzer I_{max}	A	5000
Pump displacement V_L	mL·r^{-1}	5000
Pump/motor mechanical efficiency	/	0.9
Pump/motor volumetric efficiency	/	0.95
Electrical motor/generator efficiency	/	0.9
Accumulator volume of hydro-pneumatic energy storage unit V	m^3	3000
Accumulator maximum pressure of hydro-pneumatic energy storage unit p	MPa	15
Energy storage setting value k_1	MW	14.5
Energy release setting value k_2	MW	1

4. Results and Discussion

Wind speed data for one week as shown in Figure 6 is used as the input to the simulation models. The data can be downloaded in Appendix A. The cut-in wind speed and rated wind speed are 3 m/s and 10.56 m/s, respectively. The simulation is carried out to analyse the performance of both the conventional system and the proposed system.

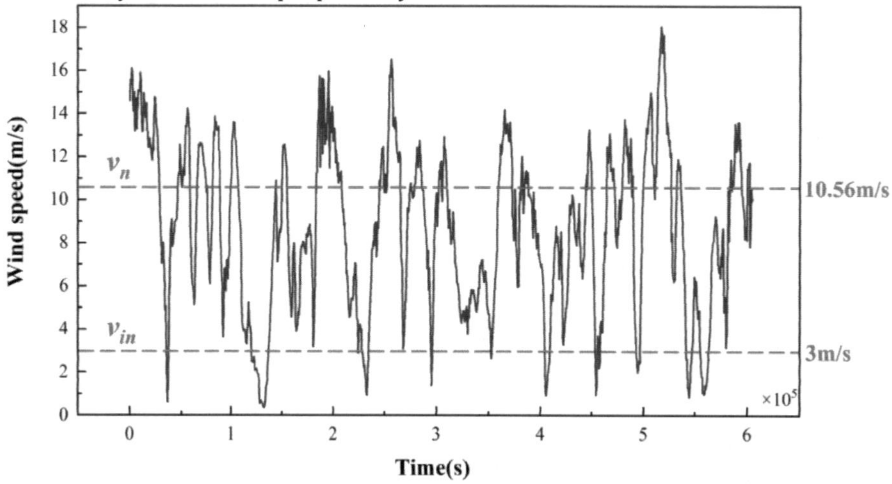

Figure 6. One week of wind speed data used as input in the numerical models.

Figure 7 illustrates the load profiles of the electrolyzers in the conventional system and the proposed system. From the local amplification of (b) and (c), it can be observed that the energy stored at the high power is well utilized in the low power part, reducing the number of on/off switching of the PEM electrolyzer in the proposed system. The hdro-pneumatic energy storage unit decreases the On/Off number of the electrolyzer. This is expected to have a favoruable impact on the electroyzer lifetime. The numericals models estimate a 83.3% reduction in the On/Off switching by intrdicuing the hydro-pneumatic energy storage system.

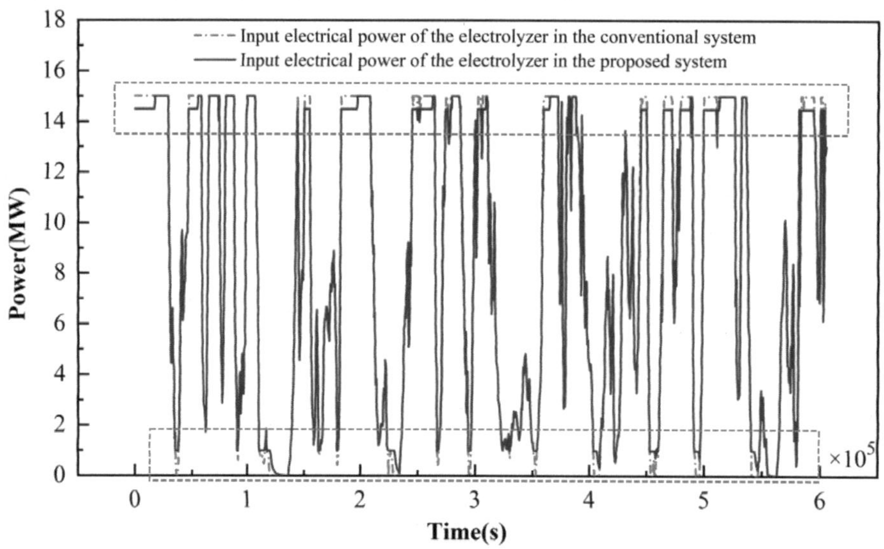

(a)

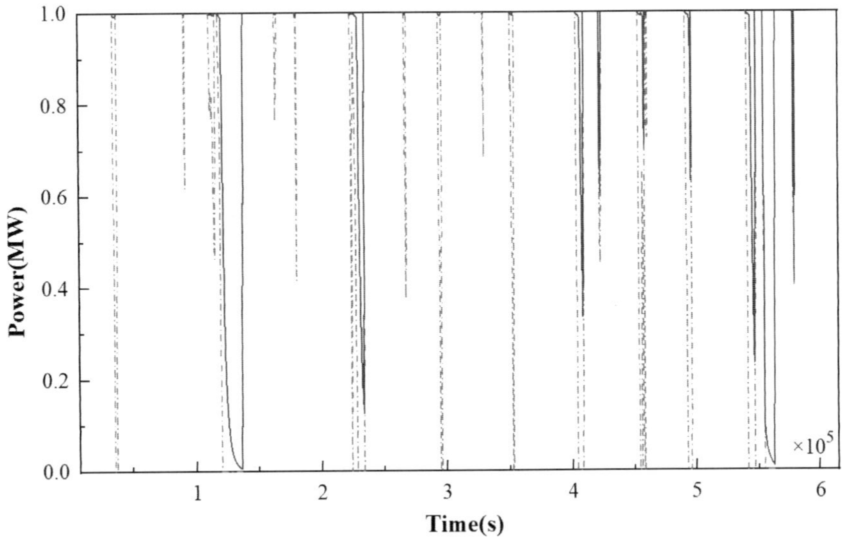

(b)

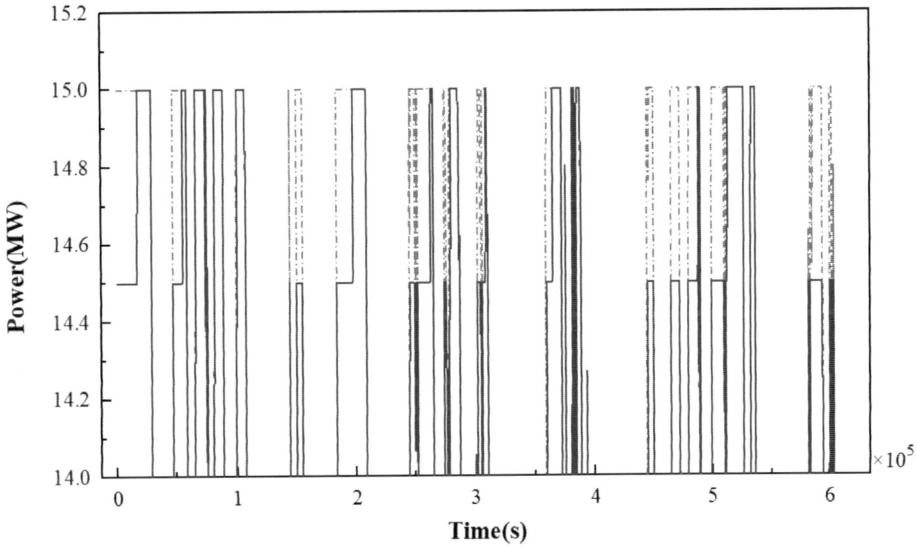

(c)

Figure 7. (a) The load profiles of PEM electrolyzers in two systems; (b) Zoomed-in view of the Low power compensation with time; (c) Zoomed-in view of the excess energy storage with time.

Figure 8 compares the instantauonaeous and cumulative values for hydrogen production of the two systems. For the week of wind data cnsidererd (Figure 6), the total hydrogen produced from the proposed system is estimated to be around 180 kg higher than that of the conventional system. This is equalivent to a percentage increase in hydrogen output of 0.85 %. Figure 8 (b) is zoomed-in view of the instantaneous hydrogen production with time. Compared with the two systems, the instantaneous hydrogen production in the proposed system is compensated because the number of electrolyzer On/Off is reduced.

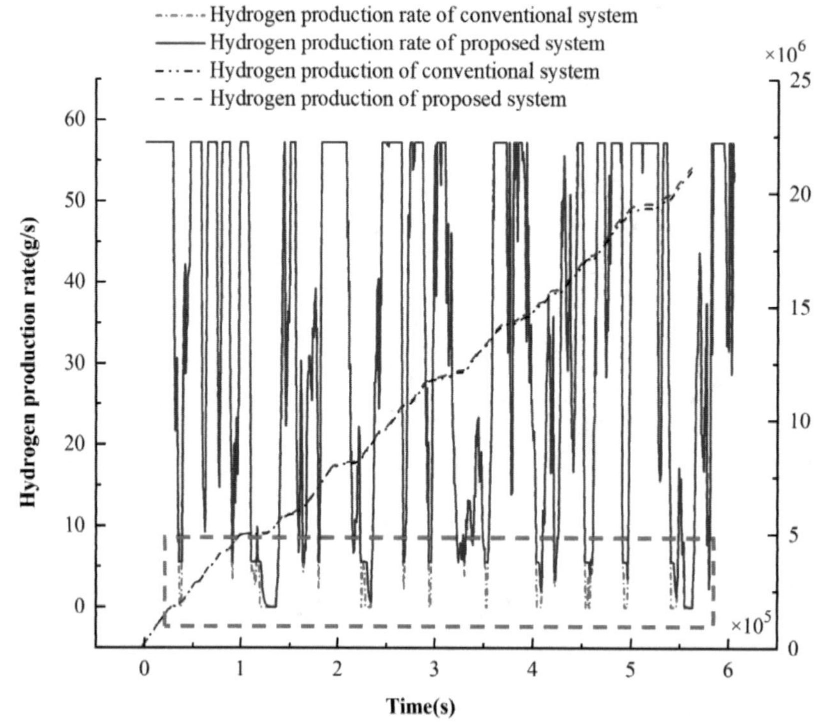

(a)

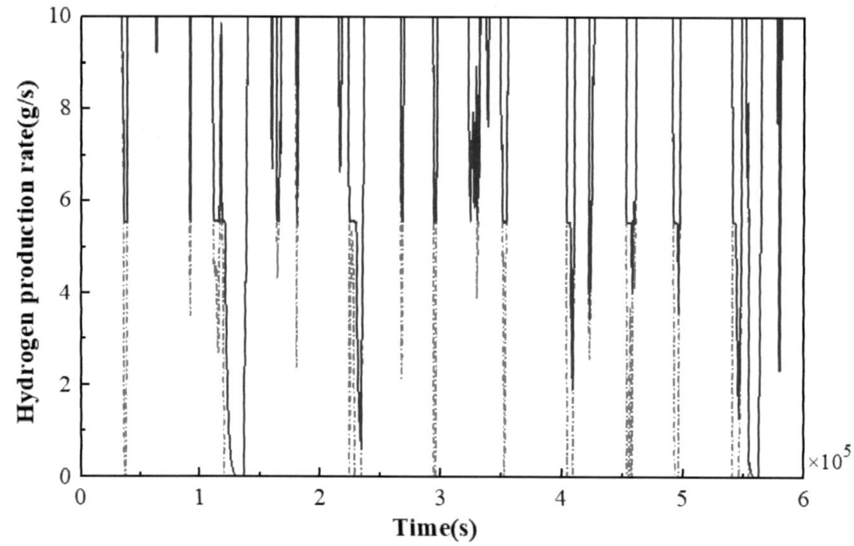

(b)

Figure 8. (a) Variation of the instantaneous hydrogen output with time and the cumulative production for the conventional and proposed systems; (b) Zoomed-in view of the instantaneous hydrogen production with time.

Figure 9 shows the efficiency and utilization rate of each part of the system. The electrical energy efficiency of the proposed system and the conventional system are 96.85% and 96.32%, respectively. The proposed system exhibits a 0.53% higher electrical energy efficiency compared to the conventional system. The energy input from wind power to the electrolyzer is not fully utilized. In the proposed system, the hydro-pneumatic system stores the excess energy while providing compensation when the wind power is low and avoiding the electrolyzer downtime. For hydro-pneumatic energy storage unit, its efficiency is 53.3%, in which the efficiency of the energy storage side is 77.21%, and the efficiency of the energy release side is 77.1%. In the hydro-pneumatic energy storage unit, the efficiency of the motor, pump and motor is taken into account to the lower energy efficiency. The electrical energy density of the unit is also calculated to be 3.28 kWh/m3. In addition, the electrical energy utilization rate of electrolyzer in the proposed system and the conventional system are 97.44 % and 96.32%, respectively.

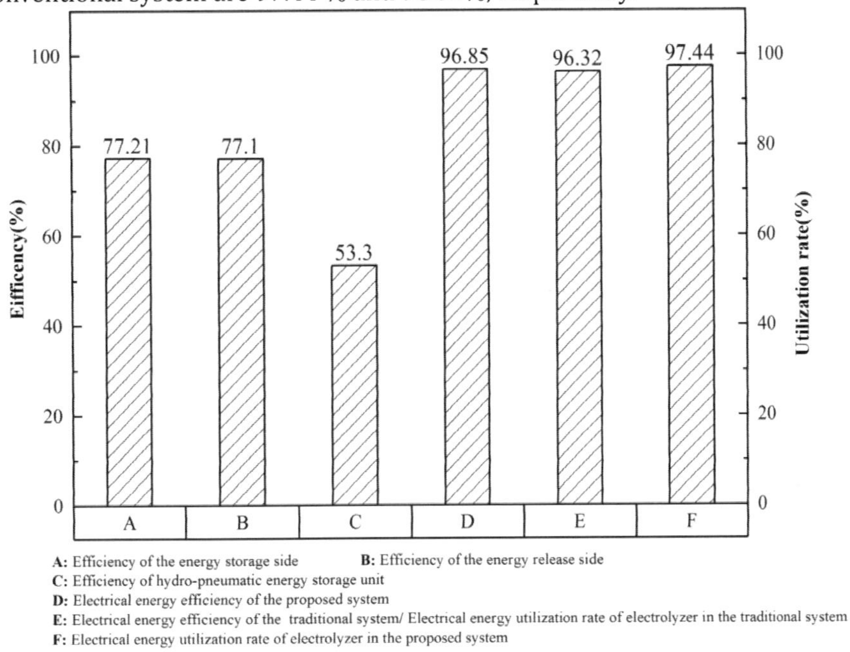

Figure 9. The efficiency and utilization rate of each part of the system.

Figure 10 shows the pressure changes of stored gases in the accumulators of hydro-pneumatic energy storage unit, subsea isobaric hydrogen storage, and conventional isochoric hydrogen storage. The pressure curve of the accumulator in the hydro-pneumatic energy storage unit can be used to characterize the energy storage and release. It can be found that the amount of energy to be compensated in different time periods is different, which is determined by the input time profiles of wind speed. The imbalance of the energy storage and release in the accumulator is also caused. In fact, the energy management in the accumulator should be also considered to realize the balance of energy storage and release in the accumulator. The detailed energy management strategies should be optimized in specific applications. In addition, Figure 10 also shows the hydrogen pressure changes in subsea isobaric hydrogen storage and conventional isochoric hydrogen storage. Due to the intermittence of ship transport, the imbalance between hydrogen production and consumption is alleviated by conventional isochoric hydrogen storage or subsea isobaric hydrogen storage storing excess hydrogen. Compared with the conventional isochoric hydrogen storage, subsea isobaric hydrogen storage can achieve constant pressure output.

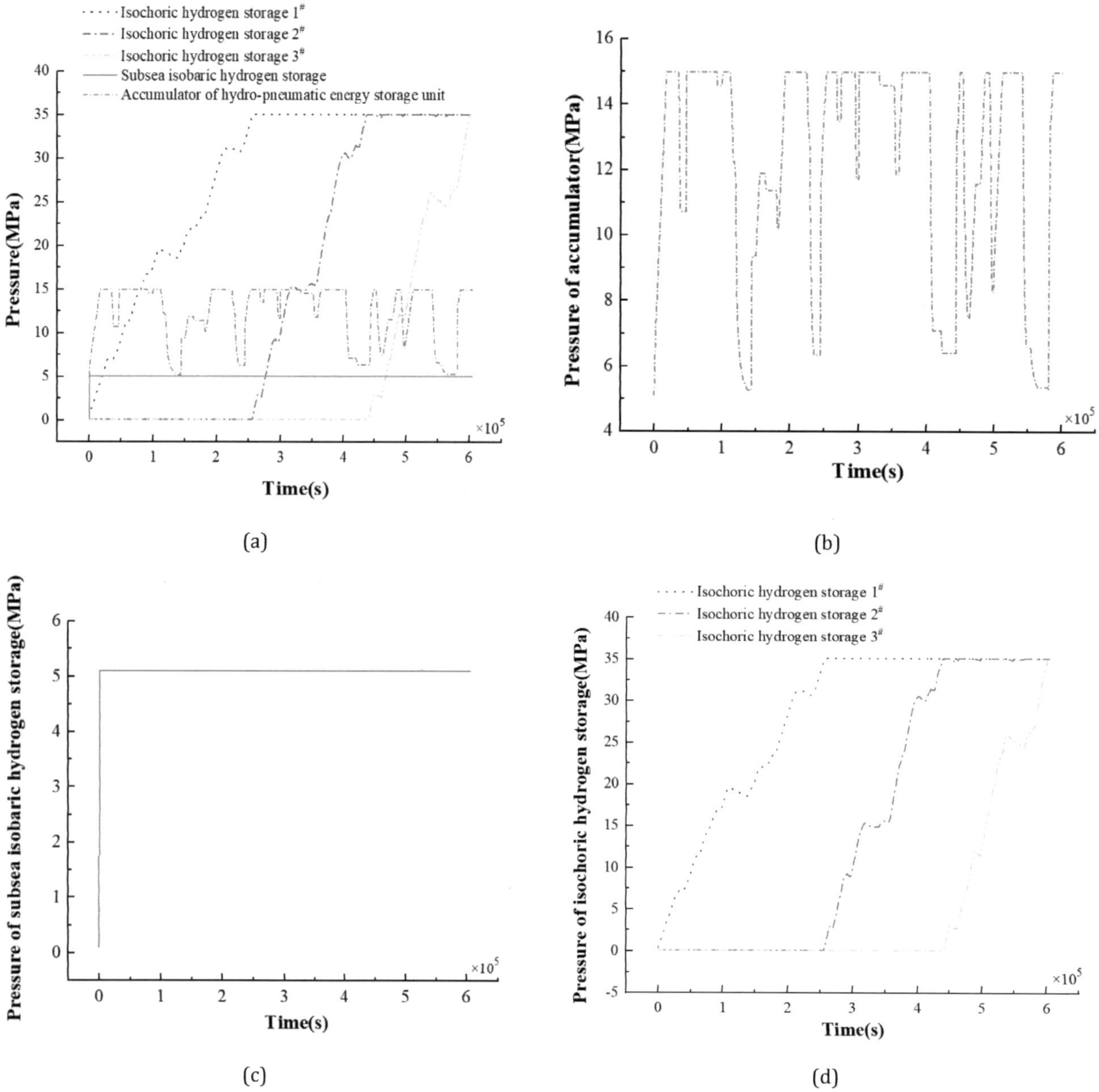

Figure 10. The pressure changes, (b) the accumulator of hydro-pneumatic energy storage unit, (c) the subsea isobaric hydrogen storage, (d) the conventional isochoric hydrogen storage.

The temperature changes in the accumulator of hydro-pneumatic energy storage unit, subsea isobaric hydrogen storage and conventional isochoric hydrogen storage is shown in Figure 11. Due to the good heat dissipation advantage of the water, the temperature changes in hydro-pneumatic energy storage unit and subsea isobaric hydrogen storage are very small, and the maximum temperature differences are 0.7 K and 6.4 K, respectively. The thermodynamic process in the two devices is close to the isothermal process. It is beneficial to reduce the energy loss and improve the efficiency of the system. For conventional isochoric hydrogen storage, its temperature changes signficantly. As a result, the heat lost is high and the thermal management is a significant challenge.

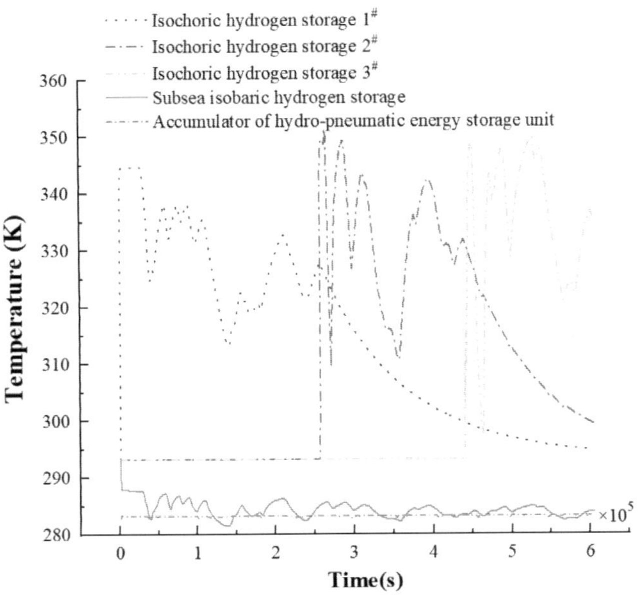

(a)

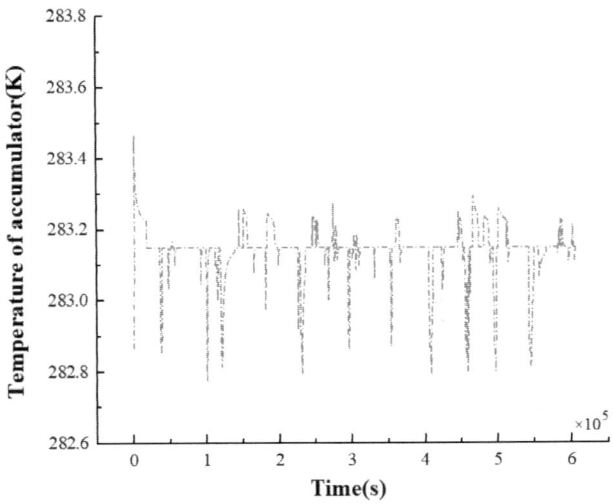

(b)

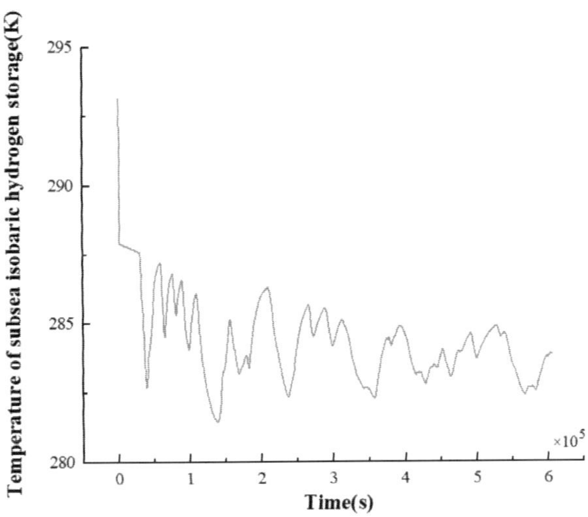

(c)

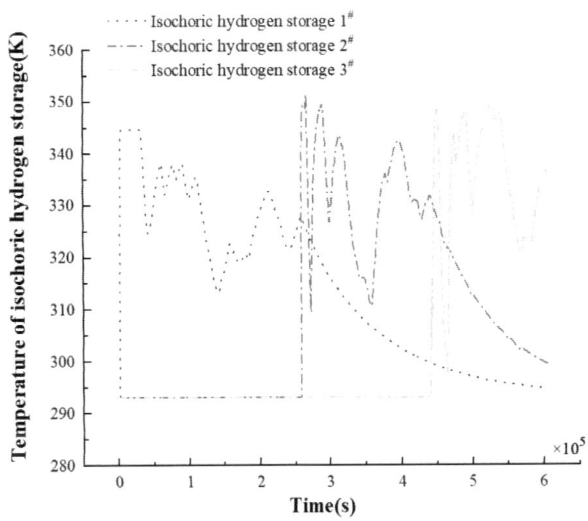

(d)

Figure 11. The temperature changes, (b) the accumulator of hydro-pneumatic energy storage unit, (c) the subsea isobaric hydrogen storage, (d) the conventional isochoric hydrogen storage.

5. Conclusions

With the rapid development of renewable energy utilization, hydrogen production with offshore wind is a promising solution for effectively harvesting potent offshore wind energy. However, the intermittency of offshore wind power poses a significant challenge to PEM electrolyzers. This study presents a decentralized floating offshore wind hydrogen production system with hydro-pneumatic energy storage and subsea isobaric hydrogen storage. A simulation analysis was carried out. The major conclusions of the study are summarized as follows.

- The hydro-pneumatic energy storage unit reduces the number of On/Off switching of the hydrogen production system by 83.3%. This will improve and extend the life of the hydrogen production system. It is also safer than battery energy storage as no significant fire hazards are present.
- For the hydro-pneumatic energy storage unit, its efficiency is 53.3%, in which the efficiency of the energy storage side is 77.21%, and the efficiency of the energy release side is 77.1%. The electrical energy density of the unit is also calculated to be 3.28 kWh/m^3.
- Compared with the conventional isochoric hydrogen storage, the subsea isobaric hydrogen storage and hydro-pneumatic energy storage unit take advantage of the good heat dissipation of water and reduces the energy loss. The maximum temperature differences are 0.7K and 6.4K, respectively.

Overall, several advantages can be achieved by integrating HPES and subsea isobaric hydrogen storage into floating offshore wind hydrogen production systems. Nevertheless, it should be noted that this is only a preliminary feasibility study with non-optimized parameters. Parametric analysis and optimization must be conducted for specific cases as part of future work. it is worth noting that the sensitivity of the efficiency of the subsea motor and pump components to operational parameter variation can illustrate opportunities for system improvement. Secondly, due to the limitations of the size of the accumulator and the difference in the amount of compensation energy, the imbalance of energy storage and energy release is also a critical point of study in the subsea hydro-pneumatic energy storage unit.

Appendix A. Supplementary data

https://download.csdn.net/download/qq_41675873/12007633?ops_request_misc=&request_id=&biz_id=103&utm_term=%E9%A3%8E%E9%80%9F2012&utm_medium=distribute.pc_search_result.none-task-download-2~all~sobaiduweb~default-3-12007633.142^v100^pc_search_result_base8&spm=1018.2226.3001.4187.4

References

[1] Robinson A, Lehmann J, Barriopedro D, Rahmstorf S and Coumou D 2021 *npj Clim. Atmos. Sci.* **4** 45
[2] Kojima H, Nagasawa K, Todoroki N, Ito Y, Matsui T and Nakajima R 2023 *Int. J. Hydrogen Energy* **48** 4572-4593
[3] Ji J, Chi Y, Yin X 2024 *Int. J. Hydrogen Energy* **62** 99-108
[4] URL https://www.iea.org/reports/net-zero-by-2050
[5] Bento N, Fontes M 2019 *RENEW SUST ENERG REV* **99** 66-82
[6] Zountouridou E I, Kiokes G C, Chakalis S Georgilakis P. S and Hatziargyriou N. D 2015 *RENEW SUST ENERG REV* **51** 433-448
[7] Giampieri A, Ling-Chin J, Roskilly A P 2024 *Int. J. Hydrogen Energy* **52** 589-617

[8] Dokhani S, Assadi M, Pollet B G 2023 *Int. J. Hydrogen Energy* **48** 9592-9608
[9] Dincer I, Zamfirescu C 2012 *Int. J. Hydrogen Energy* **37** 16266-16286
[10] Settino J, Farrugia R N, Buhagiar D and Sant, T 2022 *J. Phys.: Conf. Ser.* **2151** 012013
[11] Sayed-Ahmed H, Toldy Á I, Santasalo-Aarnio A 2024 *RENEW SUST ENERG REV* **189** 113883
[12] Majumdar A, Haas M, Elliot I, Nazari S 2023 *Int. J. Hydrogen Energy* **48** 30621-30641
[13] Wang Z, Wang H, Sant T, Zhao Z, Carriveau R, Ting D S and Xiong W 2024 *Int. J. Hydrogen Energy.* **71** 1266-1282
[14] De Siqueira L M S, Peng W 2021 *J. Energy Storage* **35** 102252
[15] Klar R, Steidl B, Aufleger M 2018 *Ocean Eng.* **165** 328-335.
[16] Wang Z, Ting D S, Carriveau R, Xiong W and Wang Z 2016 *J. Energy Storage* **5** 203-211
[17] Pimm A J, Garvey S D and de Jong M 2014 *Energy* **66** 496-508
[18] Wang H, Wang Z, Liang C, Carriveau R, Ting D S, Li P, Cen H and Xiong W 2022 *Appl. Sci.-Basel* **12** 9361
[19] Wang Z, Wang J, Cen H, Ting D S, Carriveau R and Xiong W 2021 *Ocean Eng.* **234** 109184
[20] Slocum A H, Fennell G E, Dundar G, Hodder B G, Meredith J D C and Sager M A 2013 *Proc. IEEE* **101** 906-924
[21] Coiro D and Sant T 2019 The Institution of Engineering and Technology (IET)
[22] Sant T, Buhagiar D, Farrugia R N 2018 *Ocean Eng.* **166** 232-241
[23] Borg A, Sant T, Buhagiar D, Cutajar C and Farrugia R N 2023 *J. Phys.: Conf. Ser.* **2626** 012014
[24] Astolfi M, Guandalini G, Belloli M, Hirn A, Silva P and Campanari S 2019 *Turbo Expo: Power for Land, Sea, and Air* **58608** V003T06A009
[25] Dinh V N, Leahy P, Mckeogh E, Murphy J and Cummins V 2021 *Int. J. Hydrogen Energy* **46** 24620-24631
[26] Hassanpouryouzband A, Joonaki E, Edlmann K and Haszeldine R. S 2021 *ACS Energy Lett.* **6** 2181-2186
[27] Wang H, Wang Z, Liang C, and Xiong W 2023 *9th FPM Lanzhou China* 1-9
[28] Giampieri A, Chin J. Ling, Roskilly A P 2023 *Int. J. Hydrogen Energy* **01** 346
[29] Babarit A, Gilloteaux J C, Clodic G, Duchet M, Simoneau A and Platzer M 2018 *Int. J. Hydrogen Energy* **43** 7266-7289
[30] Scolaro M, Kittner N 2022 *Int. J. Hydrogen Energy* **47** 6478 6493
[31] https://plm.sw.siemens.com/en-US/simcenter/systems-simulation/amesim Accessed on 2024-11-08
[32] URL https://cordis.europa.eu/project/id/815083/results
[33] Lydia M, Kumar S S, Selvakumar A I and Kumar G E P 2014 *RENEW SUST ENERG REV* **30** 452-460
[34] Yigit T, Selamet O F 2016 *Int. J. Hydrogen Energy* **41** 13904-13914
[35] Di J, Shen Y, Jia M and He Q 2022 *Thermal Power Generation* **51** 65-73
[36] Sharma M, Pachauri R K, Goel S K 2015 *RDCAPE* IEEE

 doi:10.1088/1742-6596/2929/1/012013

Multiscale finite element modelling of dynamic subsea power cables in bending

D M Smith[1], L S Cunningham [1] and L Chen[2]

[1]Department of Solids and Structures, The University of Manchester, Manchester, UK

[2]Department of Electrical and Electronic Engineering, The University of Manchester, Manchester, UK

E-mail: dominic.smith@manchester.ac.uk

Abstract. This paper proposes a multiscale finite element modelling strategy for dynamic offshore power cables which allows for detailed analysis of individual components. The challenge of modelling the intricate power cores within the cable is addressed by considering a separate repeated unit cell (RUC) of the power core to define its stiffness, where the power cores themselves are then discretised with beam elements in the dynamic cable model. Periodic boundary conditions allowing for geometric non-linearity are employed at both the power core and dynamic cable scale. This means the model size can be reduced to the smallest possible RUC while still permitting non-linear material behaviour. The resulting model is capable of detailed stress analysis of individual components required for fatigue life estimation. The model is validated against existing experimental data showing strong agreement. The proposed multiscale approach may therefore be used to augment existing fatigue analysis methods for dynamic offshore cables where the stress state at individual wires is required.

1 Introduction

Floating offshore renewable energy systems, such as those which harvest wave, wind or solar energy, require electricity to be transmitted to shore through dynamic subsea power cables [1]. Comprised of compacted copper conductors, various polyethylene insulation layers and helical steel armour wires, dynamic cables attach to the floating energy system and float down through the water column. The cable reaches the sea bed connecting to a static cable whereby energy is then transported to shore. Dynamic cables are therefore subjected to dynamic wave and current induced loading from the motion of the floating energy system [2], while static subsea power cables rest on the sea bed and are not subjected to the same dynamic loads. The cyclic nature of the environmental loading results in fatigue being a critical design issue [3]. As the cable is mechanically constrained by the movement of the floating platform, long-term dynamic environmental loading exposes the cable components to mechanical fatigue, particularly in the case of cyclic bending [4]. The fatigue analysis of these highly complex structures remains a challenge for the industry due to the complex geometry and associated non-linearities. Assessment of fatigue life relies on accurate predictions of the mechanical stress in the metallic components resulting from the dynamic environmental loading. Due to the complex nature of the cable, the finite element method is often chosen to analyse the stress field.

There have been few dedicated dynamic offshore cable finite element modelling approaches described in the literature. Several recent publications consider a detailed mesh of the armour wires [5, 6, 7, 8] but only consider an idealised power core geometry. By neglecting to model the individual wires of the conductor and screen layer, the stiffness of the power core may not be faithfully represented and the stress distribution cannot be studied to inform fatigue life estimations. For both umbilical models [9, 10, 11] and dynamic cable models, periodic boundary conditions have been employed to take advantage of the periodic geometry. They allow the shortest possible repeated unit cell to be modelled whilst eliminating the artificially induced stress concentrations which occur with fixed boundary conditions.

In [5, 12], a homogenisation technique was used to derive periodic boundary conditions using asymptotic expansion. This method allowed the length of the model to be reduced to one cable period, however, geometric and material non-linearity cannot be considered. A RUC approach was developed by [9] and validated by [11] to analyse flexible pipes. This method accounts for non-linearity through periodic boundary conditions which link end nodes with a central plane of helical symmetry. The drawback of this approach is two cable periods are required, doubling the computational domain. The objective of the present work is to propose a multiscale finite element modelling strategy which allows for detailed analysis of individual components. This is achieved through applying the RUC methodology outlined in [13]. The challenge of modelling the intricate power cores within the cable is addressed by considering a separate RUC of the power core to define its stiffness, where the power cores themselves are then discretised with beam elements in the dynamic cable model. Periodic boundary conditions allowing for geometric non-linearity are employed at both the power core and dynamic cable scale. This means the model size can be reduced to the smallest possible RUC while still permitting geometric and material non-linearity. The resulting model is capable of detailed stress analysis of individual components required for fatigue life estimation. The benchmark experimental and numerical study of [5] is used for validation. The radial shrinkage that occurs during the extrusion of the polyethylene layers and its resultant radial force on the below metalic layers is considered here by modelling the volume reduction using a linear thermal expansion coefficient.

2 Description of the mechanical problem

2.1 Global scale configuration

The benchmark study [5] used for this validation uses a dynamic power cable from a floating offshore wind turbine configuration. The curvature ranges used for the bending load are therefore consistent with the typical environmental loading from the lazy-wave cable configuration illustrated in Figure 1(a). At the hang-off point of the cable, a bend stiffener reduces the curvature variation at the connection point. The other end is connected to a static cable laid on the seabed. Buoyancy modules are used to maintain the lazy-wave shape around the midsection of the cable length. This allows the cable to flex when the floating platform drifts, reducing the likelihood of sudden large tensile forces developing during storm events. With this configuration, the curvature variation is most significant around the buoyancy modules where the sagging and hogging zones form the lazy wave shape.

In the validation study [5] a 4-point bend test is used to provide experimental validation. The test imposed several curvature ranges mimicking typical values seen in service around the buoyancy modules for the configuration in Figure 1(a). The 4-point configuration was chosen to ensure zero shear force between the 2 load points with a constant bending moment. The results from this study are used here for validation. This paper compares the present model's bending response with that of [5], however, tensile and torsional behaviour may still be studied with the present method.

2.2 Local description of the dynamic subsea power cable

A dynamic cable comprises extremely complex geometry and various materials. As shown in Figure 1(b), cable components can be broadly categorised by their function into three groups: the armour package, the filler and the power cores. Firstly, the armour package encompasses the various layers encircling the entire cable. In this case, its primary components are an outer layer of high-density polyethylene (HDPE), two counter-wound layers of high-strength steel armour wires and an inner layer of HDPE. Its purpose is to resist the mechanical and chemical loads exerted on the cable during its installation and operation.

Next, the filler components include all the material which fills the gaps between the armour package and the power cores. Represented in Figure 1(b) by a simple cylinder, the shape and material used for the filler can vary considerably between manufacturers. In this case, it is made with three sets of polypropylene ropes. Their purpose is to maintain the stability of the armour package structure, ensuring high axial stiffness and compliance in bending.

Lastly, the power core refers to each of the three central helical components within the cable. Their function is to transmit electricity using alternating current (AC). The central conductor comprises several layers of copper wires, where each layer is drawn through a compaction die to reduce the total cross-sectional area of the resultant conductor. This process plastically deforms each wire cross-section, influencing how the conductor responses to mechanical loading. The conductor is electrically insulated by a layer of cross-linked polyethylene (XLPE), which itself is encased by two thin semi-conductive screen layers, which are neglected in this study. A layer of copper screen wires sits above the insulation which act as the ground for the cable. Finally, a layer of protective HDPE encases the power core.

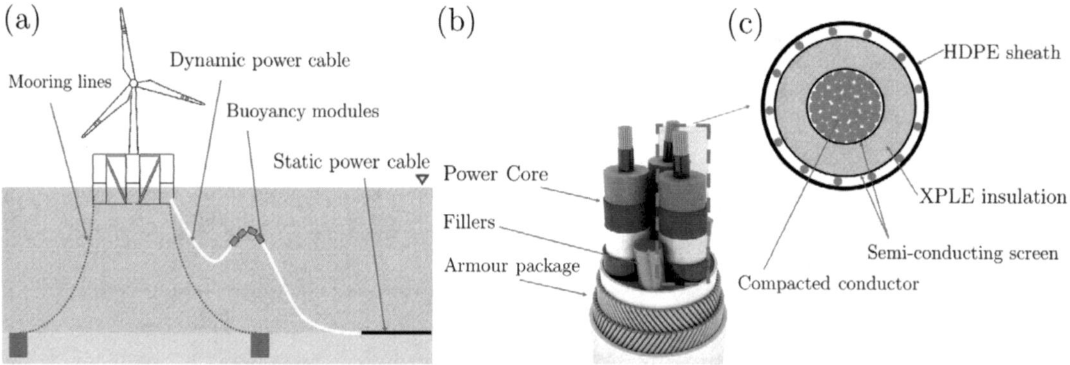

Figure 1: (a) Schematic of floating offshore wind turbine with dynamic power cable. (b) Schematic of dynamic power cable. (c) Schematic of power core cross-section.

2.3 Mechanical behaviour of cables

Above other loading conditions, bending in cables is the most strongly influenced by friction. This is due to the shear action required to resolve the tensile and compressive force between each side of the neutral axis. For a detailed discussion on simple cable bending see [14]. As a result, the magnitude of the radial forces between components greatly affects bending behaviour. Generally in the case of dynamic subsea cables these forces come from three sources: (i) tension, (ii) residual radial force from the polyethylene layers as a result of thermal contraction during the cooling which occurs post extrusion during manufacture and (iii) hydrostatic pressure. This study only considers (ii) as a source of radial force.

The resultant behaviour of cables in bending where radial forces are present is a non-linear moment-curvature response. In the initial loading phase, the cable exhibits its maximum bending stiffness, $(EI)_{max}$. Here all the wires in the cable are held together by frictional forces meaning there is no gross wire sliding. Bending loads will therefore be resisted partly by the contribution of individual wire bending and partly by the wires acting together as a unified cross-section. The transition phase begins when the stiffness changes from $(EI)_{max}$ to $(EI)_{min}$. Upon a critical curvature κ_0 the first wire begins to slip, marking the beginning of the transition phase. At the final critical curvature, κ_1, all the wires are now in the slip state and the minimum stiffness $(EI)_{min}$ is reached.

Journal of Physics: Conference Series **2929** (2024) 012013 doi:10.1088/1742-6596/2929/1/012013

3 Finite element models

Given the extremely complex geometry of dynamic offshore cables, a detailed 3D finite element model considering each individual component, even over the relatively short length of 2.5 m used in the test set up in [5], would be prohibitively computationally expensive. Therefore it is necessary to utilise the periodicity of the helical cable geometry to reduce the computational domain through periodic boundary conditions. In the benchmark study by Menard et al. [5] periodic boundary conditions were derived using a computational homogenisation technique. The main difference between the boundary condition method used by Menard et al. and the RUC method used here is the incorporation of geometrically non-linear effects. The drawback of the present method is that a model length of 2 cable periods is required, doubling the computational domain (see Section 3.1). However the inclusion of geometric non-linearity is important when considering the residual radial force from the polyethylene layers as a result of thermal contraction during the cooling which occurs post extrusion during manufacture.

The proposed RUC modelling approach is implemented in the commercially available finite element code ABAQUS Standard 2018 [15]. The implemented model is a three-dimensional implicit time integration model that accounts for geometric non-linearity that occurs due to slippage between wires in the cable. The RUC approach uses a Fortran User Subroutine to augment the capabilities of ABAQUS/Standard through a non-linear multi-point constraint (MPC) used to implement the boundary conditions necessary for the RUC approach. The model implementation is outlined in the subsequent subsections.

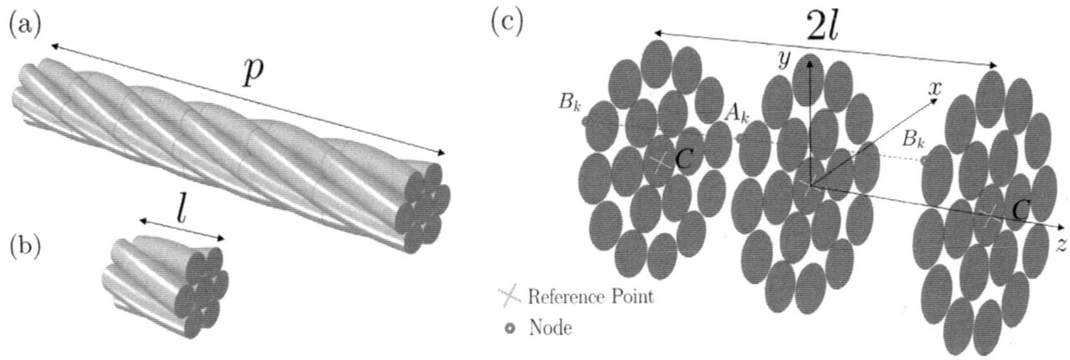

Figure 2: (a) A single-layered helical strand cable with length equal to the helix period and (b) The RUC of the cable (c) Nodes involved in periodic boundary conditions.

3.1 Periodicity of helical structures

All of the models proposed in this paper utilise the periodicity of the helical cable geometry to reduce the computational domain through periodic boundary conditions. The theoretical basis for this type of model is outlined in [13]. In essence a structure has geometric periodicity when its geometry is repeated in one or several dimensions over given lengths. The centreline of the helical components in the dynamic cable, in this case the power core, the filler and armour wires, has geometric periodicity in its axial direction over length l, where l is defined in Eq. 2. The period of a helix is well-known to be:

$$p = \frac{2\pi R}{\tan(\alpha)} \tag{1}$$

where R is the helix radius, α is the helix lay angle. In the case of the dynamic cable, helical components are grouped into layers where the the helical centrelines have a common radius R and lay angle α. The common period of a given layer j of helical components can therefore be given by:

$$l_j = \frac{p_j}{n_j} \tag{2}$$

where n_j is the number of helical components. Since the dynamic cable has multiple layers the period of each layer must be equal such that $l = l_1 = l_2 = ... = l_m$ for m layers. The cable geometry can then be assumed to be periodic in the axial direction repeating every length l. This concept is illustrated in Figure 2. Figure 2(a) shows a simple cable with central straight wire and 6 helical wires whose length is equal to the period of each individual helical wire. Figure 2(b) shows the cable reduced to the length of the period of the cable as a whole, in this case $l = \frac{p}{6}$.

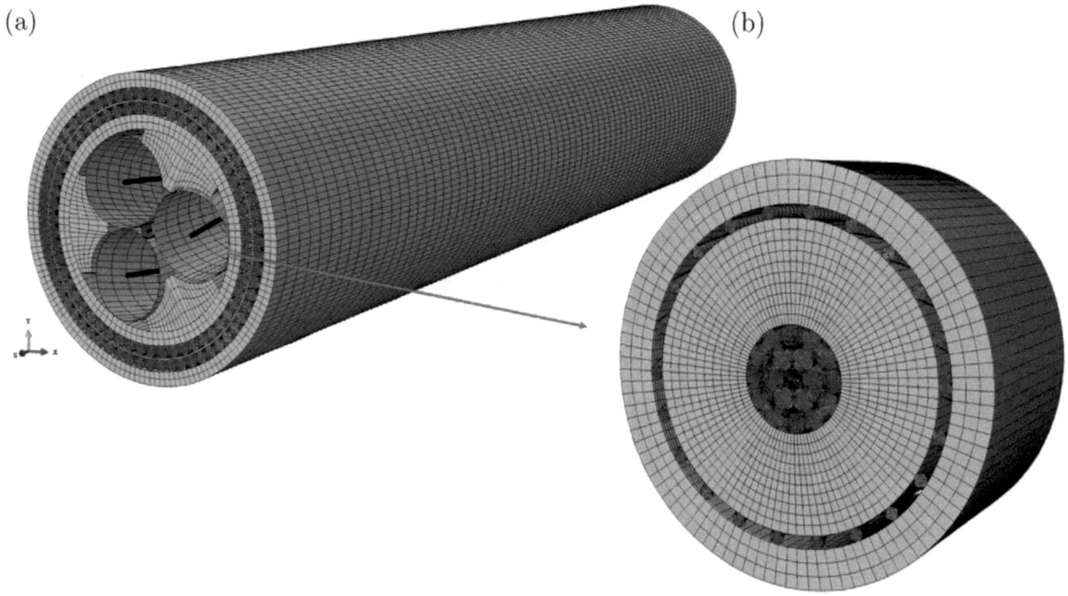

Figure 3: (a) Dynamic cable scale RUC FE model mesh (b) Power core scale RUC FE model mesh.

3.2 *Multiscale modelling with the RUC method*

One of the primary innovations proposed in this study is the use of multiscale modelling specifically for the power cores. Each power core is manufactured as a stand-alone component and laid up into the helix when the whole dynamic cable is assembled. Since each one is identical, barring any manufacturing tolerances, the proposed method uses a separate detailed model of the power core, with RUC boundary conditions to obtain the stiffness relationships in tension, bending and torsion. The dynamic cable model then uses beam elements to model power cores. The power core RUC stiffness relationships are then incorporated into the beam element section behaviour to accurately represent their non-linear stiffness while dramatically reducing the number of elements required. Both the dynamic cable model and the power core model can be seen in Figure 3(a) and (b), respectively.

The proposed multiscale methodology technically involves three scales, the first being a global model which incorporates the whole floating offshore wind turbine and dynamic cable configuration. The second and third scale being the dynamic cable scale model and the power core scale model, respectively. These two are the subject of the present study. The purpose of the global scale model would be to take the wind and wave loading on the floating platform and dynamic cable at a given site and calculate the maximum tensile forces and curvatures experienced by the cable. These values would then be used as loads for the dynamic cable scale model. There are many well-developed commercial software available for this task such as OrcaFlex or DeepLines. Since the present study uses the experimental results of Menard et al. [5] for validation, the global scale model is omitted.

Table 1: Geometrical data for each layer of the power core RUC model. From benchmark study [5].

Name	**Number**	**Material**	**Outer radius (mm)**	**Thickness (mm)**	**Lay angle (°)**
Sheath	1	HDPE	15.3	3.00	-
Screen wires	22	Copper	12.3	1.00	21.970
Insulation	1	XLPE	11.3	7.3	-
Conductor layer 2	12	Copper	4.0	1.524	11.633
Conductor layer 1	6	Copper	2.476	1.665	-11.939
Conductor central wire	1	Copper	0.853	1.706	-

Table 2: Material data for the power core and dynamic cable RUC models. From benchmark study [5].

Name	**Youngs modulus (MPa)**	**Possion's ratio**
Steel	210000	0.3
Copper	110000	0.3
HDPE	1380	0.4
PP	1300	0.4
PFR	512	0.4
XLPE	350	0.4
Central beam	100	0.4

3.3 Power core RUC model

The mesh for the power core scale model is shown in Figure 3(b). The geometric data for the power core model are shown in Table 1. The geometry of the power core is faithfully represented aside from the semi-conductive screen layers which encase the XLPE, these are omitted in the FE model. Each copper wire has a refined mesh capable of capturing the stress distribution, where a mesh density inline with the mesh convergence study carried out by [12] for helical strand cables is used. A total of 420,403 solid 8-node first order elements are used, termed C3D8 in ABAQUS. Each material is modelled as linear elastic except for the copper wires. The linear material data is shown in Table 2. For the copper wires, tensile test data was used to generate a plasticity curve to be used with an isotropic strain hardening model.

Contact is discretised using the node-to-surface method. This discretisation is chosen over the more costly surface-to-surface approach to reduce computation time since good results were obtained in the similar study by [5]. Since geometric non-linearity is being considered, large slips are expected to occur therefore finite sliding is chosen for contact tracking. The penalty method is chosen as the resolution algorithm for both normal and tangential contact definitions. For normal contact, the penalty stiffness is set to the ABAQUS default of 10 times the representative underlying element stiffness. For tangential contact, an elastic slip is permitted in the stick regime. The critical relative elastic slip is proportional to the characteristic element length in ABAQUS where the proportionality is defined by the factor β which is set to the ABAQUS default of 0.005. In the power core model, the friction coefficient for contact between the wires in the conductor is set to 0.05 to account for the water blocking paste based on the experimental study of [16]. The remainder of the contact interactions within the power core use a value of 0.15. This value is based on the experimental study by [17] for contact between polyethylene and metal.

Table 3: Geometrical data for each layer of the dynamic cable RUC model. From benchmark study [5].

Name	Number	Material	Outer radius (mm)	Thickness (mm)	Lay angle (°)
Outer sheath	1	HDPE	50.5	4.5	-
Outer armour	63	Steel	46.0	4.0	19.60
Bedding	1	PFR	42.0	0.6	-
Inner armour	55	Steel	41.4	4.0	-20.10
Inner sheath	1	HDPE	37.4	4.5	-
Power core	3	-	32.9	30.6	9.0
Central beam	1	-	1.15	2.3	-

3.4 Dynamic cable RUC model

The mesh for the dynamic cable scale model is shown in Figure 3(a). The geometric data for the power core model are shown in Table 3. The geometry of the armour package is faithfully represented, whereas the filler cross-sections are idealised as solid blocks which conform to the surfaces of the surrounding components. All components bar the power cores are meshed with solid elements having the same mesh density as used in [12]. The power cores here are meshed with 3-node Timoshenko shear flexible beam elements with quadratic interpolation, termed B32 in ABAQUS. These were found to converge faster than the linear equivalent. The stiffness of the power core beam elements is inputted manually using a non-linear general beam section in ABAQUS. This functionality allows non-linear stiffness data acquired from the power core RUC model to be used to define the stiffness of beam elements. Hence, no material definition is required for the beam elements representing the power core.

To resolve contact between the beam elements and the surrounding solid elements, 4-node surface elements, termed SFM3D4 in ABAQUS, are used to discretise the surface of the beams. These elements have no stiffness or mass. Each surface element node, therefore, is constrained to the adjacent beam element node using a rigid beam multi-point constraint. A total of 383,040 elements are used. Each material is modelled as linear elastic where the material properties are shown in Table 2. Contact is applied in the same way as with the power core model using a friction coefficient of 0.15. When resolving normal contact between the surface elements, since they have no inherent stiffness, contact penalty stiffness is manually inputted where a value of 2000 N/mm is used. This value was taken from the similar study by [11].

3.5 Thermal contraction of polyethylene layers

Another innovation in this study concerns the residual radial force from the polyethylene layers. In previous studies [5, 7, 6] the force is either omitted or idealised as a constant pressure applied to the outer layer whose magnitude is determined through sensitivity studies of the whole model with various pressures and compared to experimental results. For the first time in the published literature, this study proposes a methodology to simulate the manufacturing process by modelling the thermal contraction for each of the polyethylene layers to obtain the radial force on the metal wires below. The RUC periodic boundary conditions allow for uniform radial contraction while maintaining a constant length. Cable geometry changes significantly during this process, hence the importance of including geometric non-linear effects. One advantage of this method is the ability to model of residual force in each layer, rather than just applying a pressure to the outer layer. By separately modelling the contraction of each polyethylene layer, a more accurate distribution of internal forces will be achieved. Although the resultant pressure is not validated here directly, this is left for future work, the resultant model is validated against experimental results. In the present study the thermal expansion coefficients used for HDPE and XLPE were 0.0001 and 0.0002, respectively [18]. The HDPE and XLPE layers are assumed to be extruded at 100 degrees Celsius and be cooled down to room temperature.

3.6 Boundary conditions

The RUC periodic boundary conditions impose a set of constraints on the degrees of freedom of nodes at each end of the cable model by expressing them as a function of the nodes on the central plane. The boundary conditions are implemented in ABAQUS through non-linear user-defined Multi-Point Constraints (MPC) written in a Fortran user subroutine. Figure 2(c) illustrates the two end planes and the central plane where the total length of the model is $2l$. Each plane has a central reference point (RP) where each RP of the two end planes is referred to as C. The finite element nodes on each end plane are termed B_k for k nodes. The nodes on the central plane are termed A_k. It is essential that the corresponding nodes on each plane have identical (x, y) coordinates, details of the mesh creation method used to achieve this can be found in [13].

The user MPC eliminates the degrees of freedom of the dependent node in the constraint, expressing them as functions of the independent nodes' degrees of freedom. Each MPC in this periodic boundary condition involves three nodes; the dependent node on the respective end plane B_k, an independent node on the central plane A_k, and the respective independent control node C. The periodic constraint, originally derived by [9], is expressed as:

$$\boldsymbol{u}^B = \boldsymbol{R}(\boldsymbol{\phi}^C)(\boldsymbol{x}^B - \boldsymbol{x}^C + \boldsymbol{u}^A) + \boldsymbol{u}^C + \boldsymbol{x}^C - \boldsymbol{x}^B \tag{3}$$

where $\boldsymbol{u}^B$, $\boldsymbol{u}^A$ and $\boldsymbol{u}^C$ represent the degrees of freedom of the nodes B_k, A_k and C. The undeformed position vectors of C and B_k are represtned by $\boldsymbol{x}^C$ and $\boldsymbol{x}^B$. The rotational degrees of freedom of the control node C are denoted with $\boldsymbol{\phi}^C$, where $\boldsymbol{R}(\boldsymbol{\phi}^C)$ defines the corresponding rotation matrix. When applying Eq. 3 with solid elements, $\boldsymbol{u}^B$, $\boldsymbol{u}^A$ and $\boldsymbol{u}^C$ represent only the translational degrees of freedom. Whereas, for the beam elements in the full scale RUC model, each of $\boldsymbol{u}^B$, $\boldsymbol{u}^A$ and $\boldsymbol{u}^C$ will represent the translational and rotational degrees of freedom.

Two constraint types are required for each model, the periodic constraint described above and a constant curvature constraint to keep the three reference points on a line of constant curvature. Both MPC's are written in the nodal version of the subroutine. For the constant curvature constraint, the central reference point is coupled to the nodes A_k of a central wire or beam using a distributing coupling constraint in ABAQUS. In the power core RUC model, these nodes A_k belong to the central copper wire in the conductor. In the full scale RUC model there is no central wire in the cable geometry, therefore a central beam is added for this purpose. This beam is disctretised in the same way as the beam elements which represent the power cores, but it has a material of negligible stiffness compared to the rest of the cable, shown in Table 2. The central node of this beam acts as the node A for the constant curvature constraint. The central reference point is then constrained in all directions to avoid rigid body motions. In the present RUC approach bending loads are applied by imposing an angular rotation to the control nodes. Since the centre of the model is fixed in all directions, an additional MPC is required to impose a y displacement at each control node. This MPC ensures the three reference points follow constant curvature, which is a key assumption of the RUC approach. The y displacement of the control node u_y^C is the dependent degree of freedom:

$$u_y^C = -R_B(1 - \cos(\phi_x^C)) \tag{4}$$

where R_B is the bending radius and ϕ_x^C is the applied rotation about the x axis. For a detailed description of how both of these constraints are implemented in ABAQUS refer to [13].

3.7 Loading and simulation parameters

For the power core model, three separate loading conditions are applied in order to generate the required stiffness properties to be inputted into the beam elements for the full scale dynamic cable model: tension, bending and torsion. For the tension load, a strand axial strain of $\epsilon_z = 0.006$ is applied through setting the degree of freedom $u_z^C = \pm l\epsilon_z$ at the positive and negative control nodes, respectively, and all other degrees of freedom to zero. For the bending load, a curvature of $\kappa = 0.1$ m^{-1} is applied to the strand through applying $\phi_x^C = \pm l\kappa$ to each control node. The degree of freedom u_y^C is controlled by the constant curvature subroutine and so released for bending loads. For the torsion load, a twist angle is applied through setting $\phi_z^C = \pm 0.001$ at each control node. For the full scale dynamic cable model, only a bending load is applied.

Since individual components are not held together by fixed boundary conditions it is necessary to constrain rigid body displacements, particularly in the case of wire sliding, to prevent local instabilities from affecting the global solution procedure. In the present study viscous damping is introduced, as was done in [13], into the Newton-Raphson scheme to dissipate rigid body displacements. This viscous damping coefficient is evaluated throughout the analysis to ensure the ratio of damping energy to total energy is less than 5%, resulting in a negligible influence on the final solution.

4 Results

4.1 Power core RUC results

The stiffness of the power core in tension, bending and torsion is plotted in Figure 4(a), (b) and (c) respectively. For each plot, the theoretical stiffness, calculated using Euler-Bernoulli beam theory, for an idealised power core is also shown. This is the stiffness of the solid core idealisation which is used in several existing FE models in the literature [5, 7, 6] which is comprised in this case of a solid copper core, an XLPE insulation layer and a HDPE sheath using the same geometry as the benchmark study [5]. As can be seen in Figure 4(a), the tensile stiffness of the power core is initially marginally stiffer than the linear idealisation. This is due to the inclusion of the screen wires which replace a portion of the HDPE cross section. At $\epsilon_z = 0.0015$ the first copper wires in the conductor begin to yield and the response deviates from the linear idealisation. At $\epsilon_z = 0.005$ the screen wires begin to yield resulting in a dramatically reduced stiffness compared with the theory.

For the bending response of the power core shown in Figure 4(b) the initial stiffness $(EI)_{max}$ is almost identical to the stiffness of the idealised theory. At a critical curvature of $\kappa = 0.0001$ mm^{-1} the wires begin to slip until the final stiffness $(EI)_{min}$ is reached. At approximately $\kappa = 0.0003$ mm^{-1} the stiffness increases slightly before returning the the previous value. This can be interpreted as wires overcoming the non-circular cross-sectional shape of the wire below as they slip. The final stiffness $(EI)_{min}$ is 38% lower than the idealised bending stiffness, illustrating the benefits of including a detailed model of the power core.

Figure 4(c) shows the torsional response of the power core RUC model compared with the linear idealised theory. The power core RUC response is initially stiffer than the idealised theory due to the inclusion of the screen wires. At $\phi_z = 0.001$ the screen wires yield resulting in a response more compliant than the theory, indicating the magnitude of their influence on the response. Overall the stiffness of the power core in tension, bending and torsion varies significantly from the linear idealised theory. In all load cases, inclusion of the screen wires and accounting for each individual wire in the conductor affects the stiffness. For the case of tension and torsion, the non-linearity of the response is mainly due to material plasticity of the copper wires. In the case of bending, the non-linearity is caused by the sliding of individual wires.

The axial stress contours for the power core RUC model in tension, bending and torsion are shown in Figure 5(a), (b) and (c), respectively, at their maximum loads. In the case of tension, both the screen wires and the conductor have yielded, while the stress in the polymeric layers is far lower owing to their lower stiffness. The stress distribution across the cross-section of each wire in the conductor is approximately uniform with small variations occurring at the contact points. For the case of bending, yielding has not occurred in the majority of the cross-section. The stress distribution across each wire in the conductor shows how each wire resists bending individually. In the case of torsion, the screen wires show the highest stress and are the only metallic component to have fully yielded. For each component, stress increases radially as is expected for torsional loads. In all models the lack of stress concentrations at the edges of the model shows the merits of the periodic boundary condition approach.

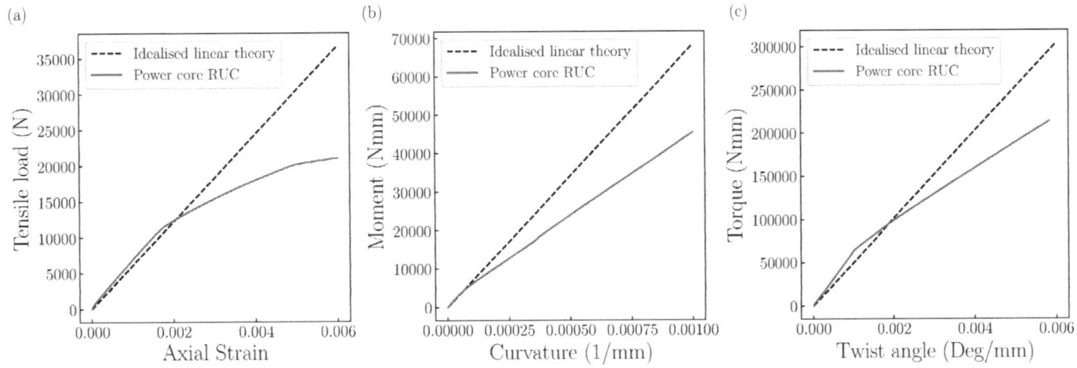

Figure 4: Stiffness relationships for the power core RUC model. (a) Tensile stiffness. (b) Flexural stiffness. (c) Torsional stiffness.

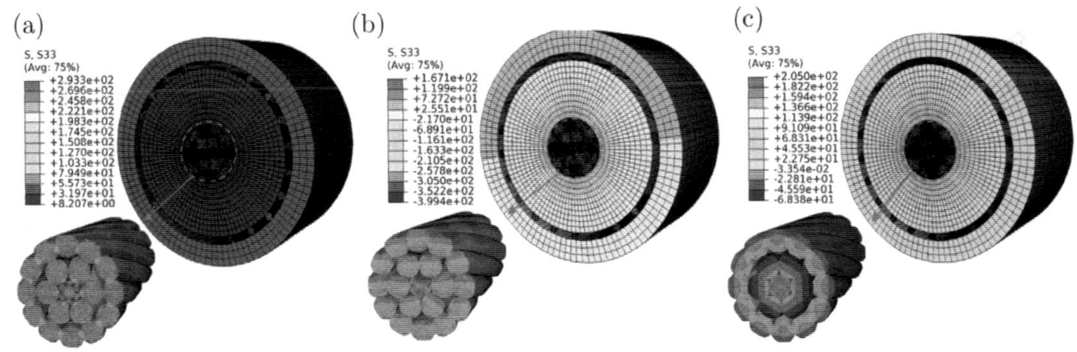

Figure 5: Axial stress (MPa) for the power core scale RUC models. (a) Tensile response. (b) Flexural response. (c) Torsional response.

4.2 Dynamic cable RUC results

The stiffness relationships for the previous section were then inputted into the general beam section behaviour in ABAQUS using the input file. A curvature of $\kappa = 0.05$ m^{-1} is then applied to the dynamic cable model and the resultant response compared with the benchmark experimental and finite element study of [5]. Figure 6(a) shows the axial stress contour at a curvature of $\kappa = 0.05$ m^{-1}. The individual steel wires of the inner layer of the armour package appear to be experiencing mainly axial loads while the wire in the outer layer shows evidence of bending behaviour. The presence of the radial contraction forces are seen in the HDPE layers of the sheath which show a very high tensile stress. Figure 6(b) plots the bending response of the dynamic cable RUC model compared with the experimental and finite element results of Menard et al. [5]. Both the finite element models are plotted from $\kappa = 0$ to 0.05 m^{-1} while for the experimental results, only the hysteresis loop is plotted since that is how the data are presented in [5]. Overall there is good agreement between the present RUC model and the benchmark model in both the initial stiffness phase and the final stiffness phase. The critical curvature of the RUC model is delayed with respect to the benchmark model, which is most likely due to the difference in friction coefficients used. When compared to the experimental results, the RUC model shows a 26% decreased stiffness in the stick phase, while the slip phase shows very good agreement.

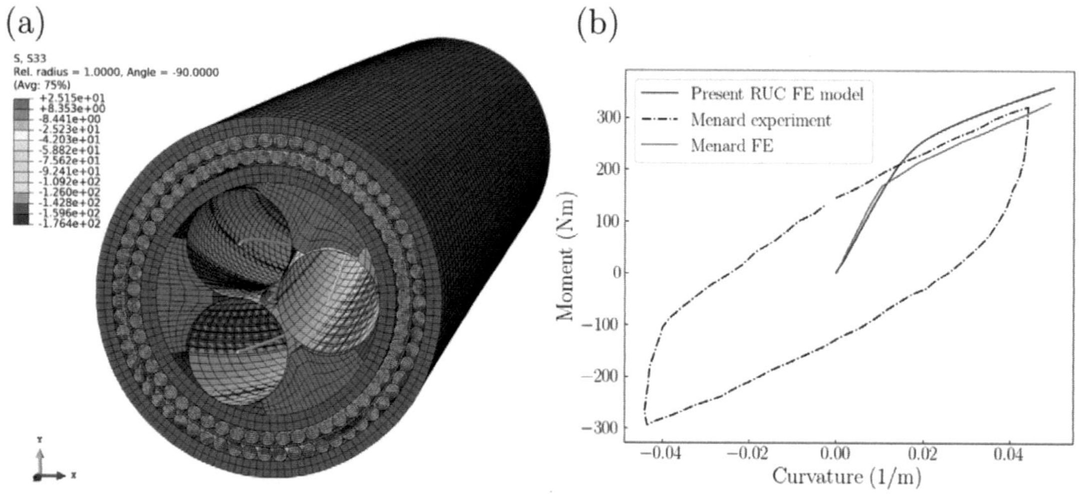

Figure 6: Dynamic cable scale RUC finite element model results (a) Axial stress contour (MPa) (b) Moment curvature relationship plotted for the present RUC model, experimental data and FE result from Menard [5].

5 Conclusion

A multiscale methodology is applied to the power cores in a dynamic offshore cable utilising the RUC method. This allows each individual wire of the conductor and screen wire layer to be considered. Fatigue analysis relies on detailed and accurate knowledge of the cyclic stress field in the component of interest. This new approach provides a way to predict the stress field in individual wires and their contact points, where previous dynamic cable models only consider an idealised geometry. While not considered in this paper, by using the methodology outlined in [13], cyclic loading may be incorporated into the multiscale dynamic cable model. Therefore, this proposed approach may be used to improve existing fatigue analysis methods for dynamic offshore cables.

When comparing the stiffness of the power core in tension, bending and torsion, to the linear theory of the idealised power core, the response deviates due to material and contact non-linearities. This is particularly important in bending as the non-linearity in the response is due to wire sliding which cannot be considered in a idealised conductor model. The final bending stiffness was shown to differ 38% when compared to the linear theory, indicating the potential errors incurred when wire sliding is omitted.

The stiffness of the power core model is programmed into a beam element which is used to model the power cores in a full-scale dynamic cable model. Using the surface element approach proposed by [5] contact is resolved between the beam elements and the rest of the model. The dynamic cable model gives a good approximation of the final bending stiffness when compared with the experimental results of [5], however, there is a difference of 26% when compared with the initial stiffness. Overall the model gives a good approximation of the dynamic cable behaviour while allowing detailed investigations of the stress profile in the power cores.

References

[1] Taninoki R, Abe K, Sukegawa T, Azuma D And Nishikawa M 2017 Dynamic cable system for floating offshore wind power generation *SEI technical review* **84**

[2] Yang S H, Ringsberg J W and Johnson E 2018 Parametric study of the dynamic motions and mechanical characteristics of power cables for wave energy converters *J Mar Sci Technol* **23** 10–29

[3] Thies P R, Johanning L and Smith G H 2012 Assessing mechanical loading regimes and fatigue life of marine power cables in marine energy applications *Proceedings of the Institution of Mechanical Engineers, Part O: Journal of Risk and Reliability* **226** 18-32

[4] Drumond G P, Pasqualino I P, Pinheiro B C and Estefen S F 2018 Pipelines, risers and umbilicals failures: A literature review *Ocean Engineering* **148** 412-425

[5] Menard F and Cartraud P 2023 A computationally efficient finite element model for the analysis of the non-linear bending behaviour of a dynamic submarine power cable *Marine Structures* **91** 103465

[6] Leroy J M, Poirette Y, Dupend N B and Caleyron F, 2017 Assessing mechanical stresses in dynamic power cables for floating offshore wind farms, *Proceedings of OMAE2017*

[7] Tjahjanto D D, Tyrberg A, and Mullins J, 2017 Bending mechanics of cable cores and fillers in a dynamic submarine cable *Proceedings of OMAE2017*

[8] Fang P, Li X, Jiang X, Hopman H and Bai Y 2024 Development of an effective modeling method for the mechanical analysis of three-core submarine power cables under tension *Engineering Structures* **317** 118632

[9] Leroy J M, Perdrizet T and Le Corre V 2010 Stress assessment in armour layers of flexible risers *29th International Conference on Ocean, Offshore and Arctic Engineering*

[10] Caleyron F, Guiton M, Leroy J M, Perdrizet T, Charliac D, Estrier P and Paumier L 2014 A multi-purpose finite element model for flexible risers studies *Proceedings of 33rd International Conference on Ocean, Offshore and Arctic Engineering* **1B**

[11] Lukassen T V, Gunnarsson E, Krenk S, Glejbøl K, Lyckegaard A and Berggreen C, 2019 Tension-bending analysis of flexible pipe by a repeated unit cell finite element model *Marine Structures* **64** 401-420

[12] Menard F and Cartraud P 2021 Solid and 3D beam finite element models for the nonlinear elastic analysis of helical strands within a computational homogenisation framework, *Computers and Structures* **25** 10675

[13] Smith D S, Cunningham L S and Chen L 2024 Efficient finite modelling of helical strand cables utilising periodicity *International Journal of Mechanical Sciences* **263** 108792

[14] Papailiou K O 1997 On the bending stiffness of transmission line conductors *IEEE Transactions on Power Delivery* **12** 4

[15] ABAQUS/Standard 2018 *Dassault Systems* https://www.3ds.com/products/simulia/abaqus, accessed 07/11/2024

[16] Karlsen S 2010 Fatigue of copper conductors for dynamic subsea power cables *Proceedings of the ASME 2010 29th International Conference on Ocean, Offshore and Arctic Engineering* **4**

[17] Dhouibi S, Boujelbene M, Leroy J M, Kharrat M, Dammak M 2013 Friction behavior of high density polyethylene (HDPE) against 304L steel: An experimental investigation of the effects of sliding direction, sliding history and sliding speed *Journal of Surfaces and Interfaces of Materials* **1** 71-76.

[18] Ashby M F Jones D R H 2013 Engineering Materials 2 (Fourth Edition) *International Series on Materials Science and Technology* Butterworth-Heinemann: Boston p. 393-404

AUTHOR INDEX